Test Item File
to accompany

ARCHITECTURAL DRAWING AND LIGHT CONSTRUCTION

Fifth Edition

Edward J. Muller
James G. Fausett
Philip A. Grau

Prepared by

Dennis F. Nieweg
Ranken Technical College

Prentice Hall
Upper Saddle River, New Jersey   Columbus, Ohio

Simon & Schuster/A Viacom Company
Upper Saddle River, New Jersey 07458

Printed in the United States of America

10 9 8 7 6 5 4 3 2 1

ISBN: 0-13-083555-2

Architectural Drawing & Light Construction . . . Muller

# of questions		
66	Chapter 1.	DRAFTING EQUIPMENT AND ITS USES
52	Chapter 2.	COMPUTER-AIDED DRAFTING AND DESIGN
51	Chapter 3.	LETTERING
51	Chapter 4.	DRAFTING EXPRESSION
26	Chapter 5.	MODULAR AND METRIC DRAFTING
58	Chapter 6.	BASIC TECHNICAL DRAWING
53	Chapter 7.	AXONOMETRIC AND OBLIQUE PICTORIAL DRAWINGS
28	Chapter 8.	FREEHAND SKETCHING
36	Chapter 9.	PERSPECTIVE DRAWINGS
16	Chapter 10.	SHADES AND SHADOWS
61	Chapter 11.	PRESENTATION DRAWINGS AND RENDERINGS
108	Chapter 12.	PRINCIPLES OF LIGHT CONSTRUCTION
30	Chapter 13.	STRUCTURAL MEMBER SELECTION
55	Chapter 14.	TYPICAL ARCHITECTURAL DETAILS
51	Chapter 15.	BASIC RESIDENTIAL PLANNING
40	Chapter 16.	BUILDING MODELS
55	Chapter 17.	WRITING SPECIFICATIONS
61	Chapter 18.	WORKING DRAWINGS OF A SMALL HOME
91	Chapter 19.	RESIDENTIAL MECHANICAL AND ELECTRICAL SYSTEMS
13	Chapter 20.	DRAWING A SMALL COMMERCIAL BUILDING

Prentice Hall: Instructor Support for Test Item Files

This hard copy test item file is just one part of Prentice Hall's comprehensive testing support service, which also includes:

1. **Prentice Hall Custom Test:** This powerful computerized testing package is designed to operate on the DOS, WINDOWS, and MACINTOSH platforms. It offers full mouse support, complete question editing capabilities, random test generation, graphics, and printing capabilities.

Prentice Hall Custom Test has a unique two-track design---*Easytest* for the novice computer user, and *Fulltest* for those who wish to write their own questions and create their own graphics.

The built-in algorithmic module enables the instructor to create thousands of questions and answers from a single question template.

In addition to traditional printing capabilities, Prentice Hall Custom Test also offers the On-Line Testing System---the most efficient, time-saving examination aid on the market. With just a few keystrokes, the instructor can administer, correct, record, and return computerized exams over a variety of networks.

Prentice Hall Custom Test is designed to assist educators in the recording and processing of results from student exams and assignments. Much more than a computerized gradebook, it combines a powerful database with analytical capabilities so the instructor can generate a full set of statistics. There is no grading system more complete or easier to use.

The Prentice Hall Custom Test is free. To order a specific Prentice Hall Custom Test title, you may contact you local rep or call our Faculty Support Services Department at 1-800-526-0485. Please identify the main text author and title.

Toll-free **technical support** is offered to all users at **1-800-550-1701.**

. For those instructors without access to a computer, we offer the popular **Prentice Hall Telephone Testing Service:** It's simple, fast, and efficient. Simply pick the questions you'd like on your test from this bank, and call the Simon & Schuster Testing Service at 1-800-550-1701; outside the U.S. and Canada, call 612-550-1705.

Identify the main text and test questions you'd like, as well as any special instructions. We will create the test (or multiple versions if you wish) and send you a master copy for duplication within 48 hours. Free to adopters for life of text use.

Chapter 1. DRAFTING EQUIPMENT AND ITS USES

1. Why are quality drafting instruments important for good work?

Answer: Quality drafting instruments are important to assure accuracy and good design.

Question Type: S
Random: Y
Difficulty Level: 1
Chapter Number: 1
Page Reference: 1

2. Why is the drawing board slightly inclined?

Answer: The drawing board is slightly inclined to create a desirable position for the drafter.

Question Type: S
Random: Y
Difficulty Level: 1
Chapter Number: 1
Page Reference: 2

3. For what purpose is a pad used under tracing paper?

Answer: A pad is used under tracing paper to protect the wood surface of the drawing board.

Question Type: S
Random: Y
Difficulty Level: 2
Chapter Number: 1
Page Reference: 3

4. Do left-handed drafters place the triangles on the T-square in the same manner as the right-handed drafter? If so, why?

Answer: Left-handed drafters place the triangle on the T-square in the same position as right-handed drafters, because the triangle must be moved from left to right to avoid smudging work that has just been completed.

Question Type: S
Random: Y
Difficulty Level: 2
Chapter Number: 1
Page Reference: 5

5. What is the advantage of transparent equipment such as triangles?

Answer: Transparent equipment such as a triangle allows the image that has been drawn to be visible through the instrument.

Question Type: S
Random: Y
Difficulty Level: 1
Chapter Number: 1
Page Reference: 5

6. Drawing pencils are available in _________ degrees of hardness.

Answer: Seventeen (17)

Question Type: F
Random: Y
Difficulty Level: 2
Chapter Number: 1
Page Reference: 4

7. Is the 2H lead harder than the 2B?

Answer: Yes, 2H lead is harder than the 2B lead.

Question Type: S
Random: Y
Difficulty Level: 2
Chapter Number: 1
Page Reference: 4

8. Define a lead holder.

Answer: A device used to hold various degrees of hardness of 2mm leads.

Question Type: S
Random: Y
Difficulty Level: 1
Chapter Number: 1
Page Reference: 4

9. Define a mechanical drafting pencil.

Answer: A device used to hold various degrees of hardness of leads and they comes in 0.3mm, 0.5mm, 0.7mm, and 0.9mm lead thicknesses.

Question Type: S
Random: Y
Difficulty Level: 1
Chapter Number: 1
Page Reference: 4

10. An adjustable triangle can only be set at every 15 degrees. True or False?

Answer: False

Question Type: T
Random: Y
Difficulty Level: 1
Chapter Number: 1
Page Reference: 6

11. The difference between the architect's scale and the civil engineer's scale is the architect's scale is in feet and inches, and the civil engineer's scale is metric. True or False?

Answer: False.

Question Type: T
Random: Y
Difficulty Level: 1
Chapter Number: 1
Page Reference: 8-9

12. List the necessary qualities of suitable tracing paper.

Answer: Desirable tracing paper should be transparent, durable, erase well, and suitable for reproduction.

Question Type: S
Random: Y
Difficulty Level: 2
Chapter Number: 1
Page Reference: 12

13. What are the most ideal lighting conditions for drafting?

Answer: Ideal lighting conditions for drafting are natural north light or soft light of the proper intensity from the upper left for right-handed drafters and vice versa for left-handed drafters. Direct sunlight should never fall on the drafting board; it is too intense.

Question Type: S
Random: Y
Difficulty Level: 2
Chapter Number: 1
Page Reference: 14

14. What is a parallel bar?

Answer: The parallel bar is a drawing tool generally constructed of wood or hard plastic and mounted to the drawing board. It is for drawing parallel lines that are horizontal.

Question Type: S
Random: Y
Difficulty Level: 2
Chapter Number: 1
Page Reference: 3

15. What is the most suitable type of lead for drafting on films ?

Answer: Plastic lead is the most suitable type of lead for drafting on films.

Question Type: S
Random: Y
Difficulty Level: 2
Chapter Number: 1
Page Reference: 4

16. What are templates and where are they used?

Answer: Templates are thin plastic sheets Out with various scaled opening. of architectural symbols. They are used on architectural drawings to represent circles, squares, rectangles, furniture, plumbing fixtures, and so on.

Question Type: S
Random: Y
Difficulty Level: 2
Chapter Number: 1
Page Reference: 10

17. How are drafting papers or films attached to the drawing surface?

Answer: Drafting paper or films are attached by drafting tape or dots which leave no marks or residue on the drawings.

Question Type: S
Random: Y
Difficulty Level: 1
Chapter Number: 1
Page Reference: 12

18. An irregular curve is one that has a number of small curvatures as well as a few long, slightly curved edges. True or False?

Answer: True

Question Type: T
Random: Y
Difficulty Level: 1
Chapter Number: 1
Page Reference: 7

19. What is the difference between the arm drafting machine and the track drafting machine?

Answer: The arm drafting machine has a pair of pinned arms and elbow. The track drafting machine has a fixed horizontal track at the top edge of the drawing board and a movable vertical track attached to the horizontal track. Arm drafting type machines are more popular with engineers and track drafting machines with architects.

Question Type: S
Random: Y
Difficulty Level: 2
Chapter Number: 1
Page Reference: 4

20. A line 3" long when measured with the scale 3/4" = 1'-0" is what length?

Answer: A 3" line drawn to the 3/4" scale is 4'-0" long.

Question Type: S
Random: Y
Difficulty Level: 3

Chapter Number: 1
Page Reference: 8

21. A line 4 1/2" long when measured with the scale 1/4" = 1'-0" is what length?

Answer: A 4 1/2" line drawn to 1/4" scale is 18'-0" long.

Question Type: S
Random: Y
Difficulty Level: 3
Chapter Number: 1
Page Reference: 8

22. A line 3" long when measured with the scale 1/2" = 1'-0" is what length?

Answer: A 3" line drawn to the 1/2" scale is 6'-0" long.

Question Type: S
Random: Y
Difficulty Level: 2
Chapter Number: 1
Page Reference: 8

23. A line 3" long when measured with the scale 1" = 10'-0" is what length?

Answer: A 3" line drawn to the 1"=10'-0" scale is 30'-0" long.

Question Type: S
Random: Y
Difficulty Level: 2
Chapter Number: 1
Page Reference: 9

24. A line 4 1/2" long when measured with the scale 1" = 10'-0" is what length?

Answer: A 4 1/2" line drawn to 1" = 20'-0" scale is 90'-0" long.

Question Type: S
Random: Y
Difficulty Level: 3
Chapter Number: 1
Page Reference: 9

25. How large should the drafting surface of a drawing board be?

Answer: Large enough to accommodate the largest drafting sheet to be used.

Question Type: S
Random: Y
Difficulty Level: 1
Chapter Number: 1
Page Reference: 2

26. What is the most desirable incline for a drafting board?

Answer: Ten degrees to fifteen degrees.

Question Type: S
Random: Y
Difficulty Level: 2
Chapter Number: 1
Page Reference: 2

27. What drafting instrument is used to draw horizontal lines?

Answer: A parallel bar.

Question Type: S
Random: Y
Difficulty Level: 2
Chapter Number: 1
Page Reference: 3

28. Name three (2) types of board covers.

1. ______________________________

2. ______________________________

Answer: 1. Vinyl   2. Paper

Question Type: F
Random: Y
Difficulty Level: 2
Chapter Number: 1
Page Reference: 3

28. Three (3) types of drafting pencils are ___________________________,

_________________ and __________________.

Answer: Wooden, lead holders and mechanical pencils

Question Type: F
Random: Y
Difficulty Level: 2
Chapter Number: 1
Page Reference: 4

30. Name three (3) types of drafting graphite-clay leads used by architectural drafters.

    1. ________________

    2. ________________

    3. ________________

Answer: 1. 2H   2. H   3. F

Question Type: F
Random: Y
Difficulty Level: 2
Chapter Number: 1
Page Reference: 4

31. A dark and heavy profile line is made with a 4H graphite-clay lead. True or False ?

Answer: False

Question Type: T
Random: Y
Difficulty Level: 2
Chapter Number: 1
Page Reference: 4

32. Describe how a parallel bar may be adjusted on the drafting board or table.

Answer: A set screw and compression plate at the top of the drawing board or table held tight against the cable keeps the bar parallel. The parallel bar may be adjusted by loosening the set screw, sliding the bar to the stops at the bottom edge of the drawing board or table and then tightening the set screw against the compression plate.

Question Type: E
Random: Y
Difficulty Level: 2
Chapter Number: 1
Page Reference: 3

33. The angle that may be set by a drafting machine is limited. True or False ?

Answer: False

Question Type: T
Random: Y
Difficulty Level: 2
Chapter Number: 1
Page Reference: 4

34. The five (5) weights of graphite-clay leads used for architectural renderings are________, ________, ________, ________, and ________.

Answer: HB, B, 2B, 3B, 4B, 5B and 6B.

Question Type: F
Random: Y
Difficulty Level: 2
Chapter Number: 1
Page Reference: 4

35. Name the four (4) weights of plastic leads and their corresponding graphite-clay weights.

1. ________ (________)
2. ________ (________)
3. ________ (________)
4. ________ (________)

Answer: 1. E-1 (F) 2. E-2 (H) 3. E-3 (2H) 4. E-4 (2H)

Question Type: F
Random: Y
Difficulty Level: 2
Chapter Number: 1
Page Reference: 4

36. Factory sharpened drafting leads do not require resharpening. True or False?

Answer: False

Question Type: T
Random: Y
Difficulty Level: 2
Chapter Number: 1
Page Reference: 5

37. What are the two (2) methods of sharpening a drafting lead?

a. __________

b. __________

Answer: 1. Sandpaper pad 2. Mechanical lead pointer

Question Type: F
Random: Y
Difficulty Level: 2
Chapter Number: 1
Page Reference: 5

38. Describe a sandpaper pad.

Answer: A sandpaper pad consists of small layers of sandpaper stapled to a wood base or handle. When the exposed sandpaper sheet is no longer cutting or sharpening, it is removed and discarded exposing a fresh strip of sandpaper. A sandpaper pad is used to sharpen wooden drafting pencils.

Question Type: E
Random: Y
Difficulty Level: 3
Chapter Number: 1
Page Reference: 1

39. The two (2) most commonly used triangles are the 30 degree and 50 degree triangles. True or False ?

Answer: False

Question Type: T
Random: Y
Difficulty Level: 2

Chapter Number: 1
Page Reference: 5

40. Why are the inside edges of triangles beveled?

Answer: To allow the drafter to pick them up easier.

Question Type: S
Random: Y
Difficulty Level: 1
Chapter Number: 1
Page Reference: 5

41. What are the advantages of an adjustable triangle over a fixed triangle?

Answer: The adjustable triangle is used to draw diagonal lines other than the 30, 45 and 60 degrees allowed by the fixed triangle. The adjustable triangle has a protractor scale adjustment from 0 to 45 degrees from the horizontal base line which allows lines of any angle to be drawn. Some adjustable triangles have scales which indicate roof slopes and stair risers.

Question Type: E
Random: Y
Difficulty Level: 3
Chapter Number: 1
Page Reference: 6

42. What is the purpose of an erasing shield?

Answer: The erasing shield is a thin, metal instrument with holes of various shapes. It is handy when eradicating small lines in confined areas without disturbing the lines to remain.

Question Type: E
Random: Y
Difficulty Level: 2
Chapter Number: 1
Page Reference: 7

43. Why are irregular or "french" curves used?

Answer: Irregular curves are those curves with nonuniform curvature that cannot be drawn with a compass or a circle template. A series of clear plastic instruments of various irregular shapes are manufactured to allow the drafting of these arcs. Adjustable curves - flexible metal cores covered with a rubber or metal ruling edge - are also manufactured for this purpose.

Question Type: E
Random: Y
Difficulty Level: 2
Chapter Number: 1
Page Reference: 7

44. Name four (4) shapes of drafting scales used by architects.

1. ____________

2. ____________

3. ____________

4. ____________

Answer: 1. Triangular 2. Two-bevel 3. Four-bevel 4. Opposite bevel

Question Type: F
Random: Y
Difficulty Level: 2
Chapter Number: 1
Page Reference: 8

45. What drafting instrument is used to draw circles and how is it used?

Answer: The friction compass is used to draw circles of minor radii and is held with the thumb and fore-finger of the drafter, leaned in the direction of travel and twirled between the fingers.

Question Type: E
Random: Y
Difficulty Level: 2
Chapter Number: 1
Page Reference: 10

46. What are the two (2) types of dividers and explain the use of each.

Answer: The two types of dividers are the basic divider and the proportional divider. The basic divider is used to transfer measurements, to divide lines into a number of parts and to transfer measurements from one drawing to another. The proportional divider is used to transfer drawings from one scale to another, to reduce or enlarge drawings and to divide a distance into a number of equal parts.

Question Type: E
Random: Y
Difficulty Level: 2
Chapter Number: 1
Page Reference: 13

47. What is the difference between an architect's scale and a civil engineer's scale?

Answer: The architect's scale is calibrated in feet and inches and is open divided. There are eleven different scales found on the architect's triangular scale. The civil engineer's scale is fully divided and is calibrated in feet and decimal parts of a foot rather than in inches.

Question Type: E
Random: Y
Difficulty Level: 2
Chapter Number: 1
Page Reference: 8-9

48. The purpose of a drafting template is ______________________________

____

____ .

Answer: The purpose of a drafting template is to provide repetitive shapes and objects on a thin plastic sheet to allow consistency in drafting and expedite drafting time.

Question Type: F
Random: Y
Difficulty Level: 2
Chapter Number: 1
Page Reference: 10

49. The equipment usually found in a good quality case or instrument package are ______________, ________________,________________, ______________, ________________ and __________________.

Answer: The equipment usually found in a good quality case or instrument package are the compass, basic dividers, ruling pen, lead holder, needle points and screw driver.

Question Type: F
Random: Y
Difficulty Level: 2
Chapter Number: 1
Page Reference: 10

50. Thumb-tacks should not be used to fasten paper to the drafting board or table. True or False?

Answer: True

Question Type: T
Random: Y
Difficulty Level: 1
Chapter Number: 1
Page Reference: 12

51. Name two (2) types of drafting media.

1. __________

2. __________

Answer: 1. Vellum 2. Drafting film

Question Type: F
Random: Y
Difficulty Level: 2
Chapter Number: 1
Page Reference: 12

52. When is thin white or light-yellow tracing paper used and what is its purpose?

Answer: Thin white or light-yellow tracing paper is used for preliminary and overlay sketching.

Question Type: E
Random: Y
Difficulty Level: 2
Chapter Number: 1
Page Reference: 12

53. List three(3) materials for points of technical ink pens.
1. ____________________
2. ____________________
3. ____________________

Answer: 1. Standard stainless steel   2. Tungsten   3. Jewel

Question Type: F
Random: Y
Difficulty Level: 2
Chapter Number: 1
Page Reference: 12

54. What is the best type of artificial light to use while drafting?

Answer: Fluorescent lighting

Question Type: S
Random: Y
Difficulty Level: 2
Chapter Number: 1
Page Reference: 14

55. Why should erasing dust bags not be used with roller type parallel bars?

Answer: The residue from the erasing dust bag will get into the rollers, pick up graphite and leave vertical dark streaks on the drawing.

Question Type: S
Random: Y
Difficulty Level: 2
Chapter Number: 1
Page Reference: 3

56. An "A" size sheet in the architectural format is ______ by _____.

Answer: 9" x 12"

Question Type: F
Random: Y
Difficulty Level: 2
Chapter Number: 1
Page Reference: 12

57. An "E" size sheet is smaller than a "B" size sheet. True or False?

Answer: False

Question Type: T
Random: Y
Difficulty Level: 1
Chapter Number: 1
Page Reference: 12

58. Define non-print leads.

Answer: Non-print leads are blue-colored leads that will not reproduce and are used to layout drawings.

Question Type: S
Random: Y
Difficulty Level: 1
Chapter Number: 1
Page Reference: 4

59. Discuss an electric eraser.

Answer: An electric eraser may be corded or cordless, and is used to hold replaceable eraser sticks.

Question Type: S
Random: Y
Difficulty Level: 1
Chapter Number: 1
Page Reference: 6

60. The five (5) common points sizes of technical pens used by architectural drafters are:
   1. ______________________
   2. ______________________
   3. ______________________

4. ______________________________
5. ______________________________

Answer: 1. 00   2. 0   3. 1   4. 2   5. 3

Question Type: F
Random: Y
Difficulty Level: 2
Chapter Number: 1
Page Reference: 13

61. Discuss the photocopy reproduction process.

Answer: This is a high-speed electronic coping process that can copy from bond paper, vellum, or film. It can copy onto these same media.

Question Type: E
Random: Y
Difficulty Level: 2
Chapter Number: 1
Page Reference: 15

MULTIPLE CHOICE: Circle the letter indicating the best choice for each question.

62. What drafting instrument is used to draw vertical lines?
    a. Scale
    b. Parallel Bar
    c. Triangle
    d. Compass
    e. None of the above

Answer: c

Question Type: M
Random: Y
Difficulty Level: 2
Chapter Number: 1
Page Reference: 5

63. The main consideration for choosing a drafting lead is:
    a. Hardness
    b. Opaqueness of Reproduction
    c. Ability to hold a point
    d. Ease of sharpening
    e. All of the above

Answer: b

Question Type: M
Random: Y
Difficulty Level: 2
Chapter Number: 1
Page Reference: 5

64. The best eraser for use on drafting film is:
    a. White plastic
    b. Art gum
    c. Rubber

Answer: a

Question Type: M
Random: Y
Difficulty Level: 2
Chapter Number: 1
Page Reference: 12

65. Some of the most used templates are:
    a. Circles
    b. Plumbing fixtures
    c. Triangles
    d. Electrical fixtures
    e. All of the above

Answer: e

Question Type: M
Random: Y
Difficulty Level: 2
Chapter Number: 1
Page Reference: 10

66. A solid continuous lightweight line is a(n) ________________.
    a. primary line
    b. hidden line
    c. center line
    d. cutting plane line
    e. dimension line

Answer: e

Question Type: M
Random: Y
Difficulty Level: 1

Chapter Number: 1
Page Reference: 15

Chapter 2. COMPUTER-AIDED DRAFTING AND DESIGN

1. Why doesn't learning to draw with a computer eliminate the need to learn to draw by hand?

Answer: Freehand and technical sketches are still required to communicate basic ideas, quickly.

Question Type: E
Random: Y
Difficulty Level: 2
Chapter Number: 2
Page Reference:

2. What is meant by the term "model" in computer aided design?

Answer: A full scaled representation of an object which can be changed and improved during the creation is termed a model.

Question Type: S
Random: Y
Difficulty Level: 2
Chapter Number: 2
Page Reference: 35

3. In what ways can computer aided design aid creativity?

Answer: Creativity can be aided by the rapid trial of ideas using CAD, in addition to visualization of spaces with walk-thrus.

Question Type: S
Random: Y
Difficulty Level: 2
Chapter Number: 2
Page Reference: 23

4. Which two major technological developments with computer hardware have fueled the rapid rise in popularity of micro computers?

Answer: Increased speed of microprocessors and reduction in cost of memory has fueled the rapid rise in micro computer popularity.

Question Type: S
Random: Y
Difficulty Level: 2
Chapter Number: 2
Page Reference: 28

5. What is the proper name for the main circuit board, which contains the microprocessor, numeric co-processor, and expansion slots?

Answer: It is known as the mother board or system board.

Question Type: S
Random: Y
Difficulty Level: 2
Chapter Number: 2
Page Reference: 29

6. Why is it important that CAD users develop good typing skills?

Answer: CAD users must develop good typing skills to be able to enter text and commands from the keyboard.

Question Type: S
Random: Y
Difficulty Level: 2
Chapter Number: 2
Page Reference: 29

7. Name a device (other than a FAX) used to transmit files electronically over telephone lines.

Answer: The modem is used to transmit files electronically over telephone lines.

Question Type: S
Random: Y
Difficulty Level: 1
Chapter Number: 2
Page Reference: 30

8. What are the major purposes of an operating system?

Answer: The main purpose of an operating system is for communications with applications software, file management (housekeeping), organization of memory, and coordination of hardware components.

Question Type: S
Random: Y
Difficulty Level: 2
Chapter Number: 2
Page Reference: 26

9. ______________ is motion along an axis from poin to point. (ie from point: x,y to point: x,y)

Answer: Linear displacement

Question Type: F
Random: Y
Difficulty Level: 2
Chapter Number: 2
Page Reference: 42

10. ______________ is motion from a known point with a distance and angle. (ie from point: x,y to point: x < angle)

Answer: Angular displacement

Question Type: F
Random: Y
Difficulty Level: 2
Chapter Number: 2
Page Reference: 42

11. In a typical CAD drawing session, the ______________________ and ______________________ commands are used most frequently.

Answer: Draw and Edit

Question Type: F
Random: Y
Difficulty Level: 2
Chapter Number: 2
Page Reference: 44

12. Why does a square one inch long on each side require the same amount of computer memory as a square one mile long on each side?

Answer: Each square is stored in the computer as four points.

Question Type: S
Random: Y
Difficulty Level: 2
Chapter Number: 2
Page Reference: 41

13. Define Hardware.

Answer: Hardware is machine components of the system.

Question Type: S
Random: Y
Difficulty Level: 1
Chapter Number: 2
Page Reference: 28

14. Define Software.

Answer: Software is the various programs that are run on the computer, such as: CAD, word processing, spreadsheets, etc.

Question Type: S
Random: Y
Difficulty Level: 1
Chapter Number: 2
Page Reference: 35

15. What are the advantages of combining drawings?

Answer: By combining drawings, parts of drawings can be re-used in other drawings.

Question Type: S
Random: Y
Difficulty Level: 1
Chapter Number: 2
Page Reference: 47

16. Discuss CAD resources available on CD-ROM.

Answer: (Answers will vary)

Question Type: E
Random: Y
Difficulty Level: 3
Chapter Number: 2
Page Reference: 37

17. Explain the difference between graphic images stored in vector format and raster format.

Answer: Graphic images are stored by their exact geometric description in a vector format and by a series of dots in raster format.

Question Type: S

Random: Y
Difficulty Level: 2
Chapter Number: 2
Page Reference: 41

18. When objects are drawn at full scale in CAD how are these scaled drawings generally produced?

Answer: The drawings are scaled as they are plotted.

Question Type: S
Random: Y
Difficulty Level: 2
Chapter Number: 2
Page Reference: 31

19. Name six (6) advantages of computer aided drafting and design over manual drafting and design.
    1. ______________________
    2. ______________________
    3. ______________________
    4. ______________________
    5. ______________________
    6. ______________________

Answer: 1. Quick, clean and convenient models of the design.
2. Reduction of office workspace.
3. Increased accuracy and precision.
4. More accurate scale of drawings.
5. Drafting flexibility, speed and efficiency.
6. Expanded possibilities of creativity in design

Question Type:
Random: Y
Difficulty Level:
Chapter Number: 2
Page Reference:

20. Discuss the difference between computer "hardware" and "software"?

Answer: Computer hardware consists of the various machines (components) of the system. A personal computer system's basic components are the central processing unit (CPU), the keyboard, the monitor and a "mouse" or "digitizer" for CADD stations. The software issues various commands for the various programs to the CPU. Various programs do various jobs; word processing, spreadsheets, CADD, etc.

Question Type: E
Random: Y
Difficulty Level: 2
Chapter Number: 2
Page Reference: 28 & 35

21. Name the basic hardware components of a personal computer CAD station.
   a. ______________
   b. ______________
   c. ______________
   d. ______________

Answer: a. Central processing unit (CPU) b. Keyboard c. Monitor d. Mouse or digitizer

Question Type: F
Random: Y
Difficulty Level: 2
Chapter Number: 2
Page Reference: 28-29

22. What are the two types of memory utilized by a personal computer?

Answer: The two types of memory utilized by a personal computer are (1) read-only-memory (ROM) which contains permanent programs used to manage the internal operation of the computer and (2) random-access-memory (RAM) which is a temporary area of memory where current work is held while being created.

Question Type: S
Random: Y
Difficulty Level: 2
Chapter Number: 2
Page Reference: 29

23. The two (2) types of magnetic disk drives are:

Answer: Diskettes and internal hard disk drives.

Question Type: S
Random: Y
Difficulty Level: 2
Chapter Number: 2
Page Reference: 29

24. What does the term "CD-ROM" identify?

Answer: The term CD-ROM identifies a compact disk - read only memory component used to store tremendous amounts of data.

Question Type: S
Random: Y
Difficulty Level: 2
Chapter Number: 2
Page Reference: 29

25. Discuss "scanning".

Answer: Scanning is the process of converting paper drawings to an electronic format that is readable by the computer. The scanner takes a picture of a paper drawing and stores it as a series of dots known as a raster format. The raster format is not as accurate as the vector format so the drawings must be converted to the vector format if a high degree of accuracy is required.

Question Type: E
Random: Y
Difficulty Level: 2
Chapter Number: 2
Page Reference: 33

26. What is a "modem" and how can it be advantageous to an architectural firm?

Answer: A modem is a device that allows for the transmission of computer data over telephone lines. With a modem, offices can connect to other computers in other locations and utilize their resources, data-bases, or share data.

Question Type: E
Random: Y
Difficulty Level: 2
Chapter Number: 2
Page Reference: 30

27. What is "DOS"?

Answer: The disk operating system (DOS) coordinates between system hardware components and applications software.

Question Type: S
Random: Y
Difficulty Level: 2
Chapter Number: 2

Page Reference: 26

28. As in analytic geometry, the ______________coordinate system is the basis for describing objects in most computer aided drafting systems.

Answer: Cartesian coordinate system

Question Type: F
Random: Y
Difficulty Level: 3
Chapter Number: 2
Page Reference: 41

29. Describe the X, Y and Z axes.

Answer: In two-dimensional form the X and Y axes project horizontally and vertically from a point known as the origin. Motion to the left or down is indicated by a negative number while motion to the right or up is indicated by a positive number. The Z axis helps to describe three-dimensional objects.

Question Type: S
Random: Y
Difficulty Level: 2
Chapter Number: 2
Page Reference: 42

30. How are commands given in the basic CAD system?

Answer: Commands in the basic CAD system can be invoked from the keyboard, screen menus, pop-down menus or the digitizing tablet with a tablet menu.

Question Type: S
Random: Y
Difficulty Level: 2
Chapter Number: 2
Page Reference: 44

31. What is a "Layer" and how does it help the professional prepare the construction documents?

Answer: A layer is an overlay on which different information may be stored. It is helpful to the professional since it allows the coordination of different disciplines by placing their respective layers on each other to determine conflicts.

Question Type: E
Random: Y
Difficulty Level: 2
Chapter Number: 2
Page Reference: 46

32. How can "blocks" or "parts" be helpful to the professional?

Answer: Blocks or parts can be helpful to the professional by allowing the copying, movement or erasure of large areas with one command, thus making the professional more efficient.

Question Type: S
Random: Y
Difficulty Level: 2
Chapter Number: 2
Page Reference: 47

33. In your opinion, what are the advantages of computer aided drafting and design to the profession?

Answer: (Answers will vary)

Question Type: E
Random: Y
Difficulty Level: 2
Chapter Number: 2
Page Reference:

34. The five system components of a CAD system are:
1. ______________________
2. ______________________
3. ______________________
4. ______________________
5. ______________________

Answer: 1. Operating system
2. Computer hardware
3. CAD software
4. Application packages
5. Customization

Question Type: F
Random: Y
Difficulty Level: 3
Chapter Number: 2

Page Reference: 24

35. In the windows environment, explain multitasking.

Answer: Multitasking is the ability to run more than one program at the same time.

Question Type: S
Random: Y
Difficulty Level: 1
Chapter Number: 2
Page Reference: 27

36. The clock speed in a CPU is measured in seconds. True or False?

Answer: False

Question Type: T
Random: Y
Difficulty Level: 2
Chapter Number: 2
Page Reference: 29

37. Computer screen resolution is measured by Dots Per Inch. True or False?

Answer: True

Question Type: T
Random: Y
Difficulty Level: 1
Chapter Number: 2
Page Reference: 29

38. Digital cameras use a special high resolution film to record photos. True or False?

Answer: False

Question Type: T
Random: Y
Difficulty Level: 2
Chapter Number: 2
Page Reference: 31

39. Except for pen/pencil plotters, resolution of the output is measured in ____________________.

Answer: DPI (dots per inch)

Question Type: F
Random: Y
Difficulty Level: 2
Chapter Number: 2
Page Reference: 31

40. Notebook computers are connected to external monitors, keyboards, and networks using a ____________________________.

Answer: Docking station

Question Type: F
Random: Y
Difficulty Level: 2
Chapter Number: 2
Page Reference: 34

41. Discuss how the internet can benefit an architectural firm.

Answer: (Answers will vary)

Question Type: E
Random: Y
Difficulty Level: 2
Chapter Number: 2
Page Reference: 38

42. Three types of CAD menus are:
    1. _________________
    2. _________________
    3. _________________

Answer: 1. Pull down 2. Tablet 3. Dialog boxes

Question Type: F
Random: Y
Difficulty Level: 2
Chapter Number: 2
Page Reference: 43

43. ________________ enables the computer to sort vectors when plotting to become more efficient.

Answer: "Fuzzy logic"

Question Type: F
Random: Y
Difficulty Level: 3
Chapter Number: 2
Page Reference: 32

44. Define "pull down" menus.

Answer: "Pull down" menus are "pulled down" from the main menu bar.

Question Type: S
Random: Y
Difficulty Level: 2
Chapter Number: 2
Page Reference: 43

45. Define "tablet" menus.

Answer: "Tablet" menus are stiff menu cards attached to a digitizing tablet and contain icons for commands. They also have a pointing area representing the screen.

Question Type: S
Random: Y
Difficulty Level: 2
Chapter Number: 2
Page Reference: 43

46. Define a "Dialog box".

Answer: Dialog boxes are boxes that display on the screen and contain values which may be changed.

Question Type: S
Random: Y
Difficulty Level: 2
Chapter Number: 2
Page Reference: 43

47. Two display commands to help move around on the screen are ______ and ____________________.

Answer: Pan and Zoom

Question Type: F

Random: Y
Difficulty Level: 1
Chapter Number: 2
Page Reference: 46

48. Discuss the meaning of "Third-party" software.

Answer: Third-party software is the customization packages that work with the CAD program. They are usually developed by outside companies (third-parties).

Question Type: E
Random: Y
Difficulty Level: 2
Chapter Number: 2
Page Reference: 47

49. Discuss the propose of the "Inquire" command.

Answer: The inquire command enables the user to obtain details from the drawing database.

Question Type: E
Random: Y
Difficulty Level: 2
Chapter Number: 2
Page Reference: 45

50. ________________ software packages are developed for specific CAD design disciplines such as: HVAC, plumbing, electrical, etc.

Answer: Application

Question Type: F
Random: Y
Difficulty Level: 2
Chapter Number: 2
Page Reference: 48

MULTIPLE CHOICE: Circle the letter indicating the best choice for each question.

51. What is the recommended minimum size of a monitor for a CAD station?
    a. 11 inches diagonally
    b. 13 inches diagonally

c. 15 inches diagonally
e. 19 inches diagonally

Answer: d

Question Type: M
Random: Y
Difficulty Level: 2
Chapter Number: 2
Page Reference: 29

52. What is the ideal size for a CAD station monitor?
    a. 11 inches diagonally
    b. 13 inches diagonally
    c. 15 inches diagonally
    d. 19 inches diagonally
    e. 21 inches diagonally

Answer: e

Question Type: M
Random: Y
Difficulty Level: 2
Chapter Number: 2
Page Reference: 29

Chapter 3. LETTERING

1. What are the distinguishing characteristics of the roman alphabet?

Answer: The roman alphabet has grace and dignity. Thick and thin-width strokes and serifs are its distinguishing features.

Question Type: S
Random: Y
Difficulty Level: 2
Chapter Number: 3
Page Reference: 53

2. Why has the gothic alphabet been the most adaptable to technical lettering?

Answer: The gothic alphabet has been the most adaptable to technical lettering because it is simple to execute and easy to read.

Question Type: E
Random: Y
Difficulty Level: 2
Chapter Number: 3
Page Reference: 53

3. Serifs are the spurs or "boots" forming the ends of the strokes which can be used on gothic letters as well as roman. True or False?

Answer: True

Question Type: T
Random: Y
Difficulty Level: 1
Chapter Number: 3
Page Reference: 53

4. Describe the meaning of lowercase lettering.

Answer: "Lowercase" lettering is a smaller rounder version of the strokes which can be used on gothic letters as well as roman.

Question Type: S
Random: Y
Difficulty Level: 1
Chapter Number: 3

Page Reference: 54

5. Which letter is the widest in the alphabet?

Answer: The letter "W" is the widest in the alphabet.

Question Type: S
Random: Y
Difficulty Level: 1
Chapter Number: 3
Page Reference: 54

6. Why are guide lines needed for lettering on working drawings?

Answer: It is nearly impossible to maintain straight lettering without guidelines. Use them like the experts do.

Question Type: E
Random: Y
Difficulty Level: 1
Chapter Number: 3
Page Reference: 56

7. What is meant by a boldface alphabet?

Answer: The boldface alphabet has heavier and thicker letters. It is formed with a thick point instrument.

Question Type: E
Random: Y
Difficulty Level: 1
Chapter Number: 3
Page Reference: 61

8. List several methods of giving titles emphasis.

Answer: Emphasis is given to titles by the use of extended lettering, thicker strokes, and/or underlining.

Question Type: S
Random: Y
Difficulty Level: 2
Chapter Number: 3
Page Reference: 59

9. What is meant by architectural style in lettering?

Answer: Architectural style is each drafters individualized lettering style which he or she reproduces with speed and consistency and is generally a variation of several styles.

Question Type: S
Random: Y
Difficulty Level: 1
Chapter Number: 3
Page Reference: 63

10. The title block is usually placed in the lower right corner of the sheet. True or False?

Answer: True

Question Type: T
Random: Y
Difficulty Level: 1
Chapter Number: 3
Page Reference: 62

11. Which of the three styles of architectural lettering illustrated in this chapter would be most appropriate for notes on an architectural working drawing?

Answer: The triangle style lettering would be most appropriate for notes on working drawings.

Question Type: S
Random: Y
Difficulty Level: 2
Chapter Number: 3
Page Reference: 67

12. What is a lettering guide?

Answer: Lettering guides are used for making uniform guide lines for lettering.

Question Type: S
Random: Y
Difficulty Level: 2
Chapter Number: 3
Page Reference: 56

13. What are lettering machines and where are they used?

Answer: Prints letters on pressure sensitive tape which are then usually used for titles.

Question Type: S
Random: Y
Difficulty Level: 1
Chapter Number: 3
Page Reference: 62

14. Why are letters not spaced equally apart in each word?

Answer: Because each letter has a different profile and width; therefore, spacing of characters within a word cannot be equal.

Question Type: E
Random: Y
Difficulty Level: 2
Chapter Number: 3
Page Reference: 58

15. When would it be appropriate to use the bold freehand letters that are executed with broad and fine-tip felt pens?

Answer: Bold freehand letters are usually used on preliminary sketches.

Question Type: E
Random: Y
Difficulty Level: 2
Chapter Number: 3
Page Reference:

16. How much space should be allowed between each word when lettering?

Answer: A good rule of thumb would be to make the space between words twice the area between letters within the word.

Question Type: S
Random: Y
Difficulty Level: 2
Chapter Number: 3
Page Reference: 59

17. Architects generally use vertical stroke letters. What type of strokes are generally used by engineers?

Answer: Engineers commonly use inclined stroke letters.

Question Type: S
Random: Y
Difficulty Level: 1
Chapter Number: 3
Page Reference:

18. What is the best method for developing perfection in lettering?

Answer: There is no substitute for diligent practice in developing perfection in lettering.

Question Type: E
Random: Y
Difficulty Level: 2
Chapter Number: 3
Page Reference: 63

19. What are the four (4) lettering classifications?

1. __________

2. __________

3. __________

4. __________

Answer: 1. Roman 2. Gothic 3. Script 4. Text

Question Type: F
Random: Y
Difficulty Level: 2
Chapter Number: 3
Page Reference: 53

20. What are serifs and on which lettering classification can they usually be found?

Answer: Serifs are the spurs or "boots" forming the ends of the strokes and are usually found on gothic and roman letters.

Question Type: E
Random: Y

Difficulty Level: 2
Chapter Number: 3
Page Reference: 55

21. Explain the difference between upper case and lower case letters.

Answer: Upper case letters are capital letters. Lower case are smaller, more rounded letters.

Question Type: E
Random: Y
Difficulty Level: 2
Chapter Number: 3
Page Reference: 55

22. What is the usual height of the lower case letter relative to the upper case letter?

Answer: The height of the lower case letter is usually two-thirds (2/3) of the height of the upper case letter with ascenders extending to the capital line and descenders dropping the same amount below.

Question Type: S
Random: Y
Difficulty Level: 2
Chapter Number: 3
Page Reference: 55

23. The vertical, single stroke alphabet is the basic style lettering used by drafters. True or False?

Answer: True

Question Type: T
Random: Y
Difficulty Level: 2
Chapter Number: 3
Page Reference: 54

24. What is the weight of the lead normally used for lettering on paper?

Answer: "H" weight lead is normally used for lettering on paper. However, the beginning drafter may want to use the "F" lead due to its darker imprint with less pressure.

Question Type: S
Random: Y
Difficulty Level: 2
Chapter Number: 3
Page Reference: 55

25. What is the weight of the lead normally used for lettering on polyester film?

Answer: "E-1" and "E-2" leads are normally used for lettering on polyester film.

Question Type: S
Random: Y
Difficulty Level: 3
Chapter Number: 3
Page Reference: 55

26. What is the drafting tool that will assure uniform height in letters?

Answer: A lettering guide or triangle will aid in maintaining uniform lettering heights.

Question Type: S
Random: Y
Difficulty Level: 2
Chapter Number: 3
Page Reference: 56

27. On working drawings at a scale of 1/4" = 1'-0", a lettering height of 1/8" should be used for minor titles and 3/32" for small notes. True or False?

Answer: False

Question Type: T
Random: Y
Difficulty Level: 2
Chapter Number: 3
Page Reference: 57

28. Spaces between letters should be nearly identical in area. True or False?

Answer: True

Question Type: T
Random: Y
Difficulty Level: 1
Chapter Number: 3
Page Reference: 58

29. Discuss how to optically judge letter and word spacing.

Answer: The observer should squint their eyes and evaluate gray tones throughout the lettering for uniformity.

Question Type: E
Random: Y
Difficulty Level: 2
Chapter Number: 3
Page Reference: 58

30. ________________, __________________, and __________________ are the three basic widths of letters.

Answer: Condensed, standard and extended.

Question Type: F
Random: Y
Difficulty Level: 2
Chapter Number: 3
Page Reference: 58

31. Why is clear, legible lettering necessary in drafting?

Answer: Drafting is a means of communication between people. If the communication cannot be read, the second party cannot understand the instructions.

Question Type: S
Random: Y
Difficulty Level: 2
Chapter Number: 3
Page Reference: 53

32. Why is the spacing of letters within a word and words within sentences important?

Answer: Poor spacing of letters often result in awkward appearance or difficult legibility. Spaces between words should be adequate to make word divisions obvious.

Question Type: S
Random: Y
Difficulty Level: 2
Chapter Number: 3
Page Reference: 58

33. When are inclined letters used and why?

Answer: Inclined letters produce contrast and variety in a drawing and may be used for minor notes and items to be emphasized.

Question Type: S
Random: Y
Difficulty Level: 2
Chapter Number: 3
Page Reference: 59

34. Name three (3) methods of providing importance to a title on a drawing.

1. ______________

2. ______________

3. ______________

Answer: 1. Extended lettering 2. Thicker strokes 3. Underlining

Question Type: F
Random: Y
Difficulty Level: 2
Chapter Number: 3
Page Reference: 59

35. Describe a method of placing a symmetrical title on a drawing.

Answer: Count the number of letters including the space between words to find the exact center. Then, starting at the desired centerline, complete the last half of the title. Complete the first half of the title by working from the center toward the beginning. 2. Lay out the entire title on a separate piece of paper. Using this as a guide, trace the title onto the final drawing.

Question Type: E
Random: Y
Difficulty Level: 2

Chapter Number: 3
Page Reference: 60

36. Name two (2) common types of lettering machines.

1. ______________________________

2. ______________________________

Answer: 1. Manually operated machines which print the letters on polyester adhesive-backed tape.
2. Word processors which transfer the letters to a "sticky-back" polyester sheet.

Question Type: F
Random: Y
Difficulty Level: 2
Chapter Number: 3
Page Reference: 62

37. What nine (9) pieces of information is usually included in a completed title block?

1. ______________________________

2. ______________________________

3. ______________________________

4. ______________________________

5. ______________________________

6. ______________________________

7. ______________________________

8. ______________________________

9. ______________________________

Answer: 1. Name and location of the structure
2. Name and location of the owner or client
3. Name and address of the architect
4. Date the drawing is completed
5. Date and type of revisions to the drawing, if any
6. Initials of the drafter
7. Initials of the checker

8. Architect's or engineer's seal
9. Identification number of the sheet.

Question Type: F
Random: Y
Difficulty Level: 3
Chapter Number: 3
Page Reference: 63

38. Name in order the letter sizes on a drawing, starting with the largest size.

a. ______________________________

b. ______________________________

c. ______________________________

Answer: a: Titles b: Sub-titles c: Notes and dimensions

Question Type: F
Random: Y
Difficulty Level: 2
Chapter Number: 3
Page Reference: 57

39. All letters are equally spaced. True or False?

Answer: False

Question Type: T
Random: Y
Difficulty Level: 2
Chapter Number: 3
Page Reference: 58

40. The spacing of numbers has different rules than the spacing of letters. True or False?

Answer: False

Question Type: T
Random: Y
Difficulty Level: 2
Chapter Number: 3
Page Reference: 58

41. What is the advantage of using condensed letters on a drawing?

Answer: The advantage of using condensed letters on a drawing is the ability to place more information on a smaller area of the drawing.

Question Type: S
Random: Y
Difficulty Level: 2
Chapter Number: 3
Page Reference: 58

42. Where would extended lettering be used?

Answer: Extended lettering is appropriate when a word or title must identify a large area or a group of drawings.

Question Type: S
Random: Y
Difficulty Level: 2
Chapter Number: 3
Page Reference: 58

43. In lettering the "W", the third and fourth strokes should be parallel to the first and second strokes. True or False?

Answer: True

Question Type: T
Random: Y
Difficulty Level: 2
Chapter Number: 3
Page Reference: 54

44. In the letter "M" the center V should terminate above the bottom guide line. True or False?

Answer: False

Question Type: T
Random: Y
Difficulty Level: 2
Chapter Number: 3
Page Reference: 54

45. Engineers generally use inclined strokes in their lettering. True or False?

Answer: True

Question Type: T
Random: Y
Difficulty Level: 2
Chapter Number: 3
Page Reference: 59

46. Where is the title block usually placed on a working drawing?

Answer: The title block is usually placed in the lower right hand corner or along the right hand edge of the drawing sheet.

Question Type: S
Random: Y
Difficulty Level: 2
Chapter Number: 3
Page Reference: 62

47. Name five (5) techniques to improve lettering.

1. ______________________

2. ______________________

3. ______________________

4. ______________________

5. ______________________

Answer:
1. Use guidelines
2. Choose one style of lettering
3. Execute bold and clear strokes in forming letters.
4. Do not draw letters; print letters with a quick, sure stroke.
5. Practice visually spacing the distance between letters and words.

Question Type: F
Random: Y
Difficulty Level: 2
Chapter Number: 3
Page Reference: chapter

48. Another name for the "text" lettering style is ______________.

Answer: Old English

Question Type: F
Random: Y
Difficulty Level: 2
Chapter Number: 3
Page Reference: 55

49. One disadvantage of CADD drawings is the lack of architectural lettering styles. True or False?

Answer: False

Question Type: T
Random: Y
Difficulty Level: 1
Chapter Number: 3
Page Reference: 68

MULTIPLE CHOICE: Circle the letter indicating the best choice for each question.

50. Which of the following type-faces is most widely used for architectural applications throughout the United States and Europe?
    a. Souvenir
    b. Gothic
    c. Helvetica
    d. Optima
    e. Roman

Answer: c

Question Type: M
Random: Y
Difficulty Level: 2
Chapter Number: 3
Page Reference: 61

51. What height letters would be appropriate for the title "NORTH ELEVATION" on architectural working drawings:
    a. 1/16"
    b. 1/4"
    c. 3/4"
    d. 1"

Answer: a. 1/4"

Question Type: M
Random: Y
Difficulty Level: 2
Chapter Number: 3
Page Reference: 56

Chapter 4. DRAFTING EXPRESSION

1. What do we mean by contrast in linework?

Answer: Contrast in linework is achieved by line width and degree of darkness. Emphasis is placed on an object when the lines forming the objects are drawn dark and bold. Less important line. are drawn narrow and light in value.

Question Type: E
Random: Y
Difficulty Level: 2
Chapter Number: 4
Page Reference: 71

2. How may broken lines or dashed lines be varied to represent various features on a drawing?

Answer: Broken lines may be varied in length and width to represent various features on a drawing. Short broken lines may represent a hidden surface above or below the surface. Long broken lines may indicate the outline of a footing on a foundation plan or electrical wiring on an electrical plan.

Question Type: E
Random: Y
Difficulty Level: 2
Chapter Number: 4
Page Reference: 71

3. What is a typical architectural characteristic of line intersections at corners?

Answer: Line intersections should meet at corners, however, a slight overlap of lines at corners is acceptable.

Question Type: S
Random: Y
Difficulty Level: 1
Chapter Number: 4
Page Reference: 73

4. From what directions on a drawing are dimensions made to read?

Answer: Horizontal dimensions should be placed above the dimension line. Vertical dimensions should be placed to the left of the dimension line.

Question Type: S
Random: Y
Difficulty Level: 2
Chapter Number: 4
Page Reference: 74

5. Why is a series of dimensions best placed along a continuous line on a floor plan?

Answer: Dimensions are easy to locate and easy to read if they are placed along a continuous line on a floor plan.

Question Type: E
Random: Y
Difficulty Level: 2
Chapter Number: 4
Page Reference: 75

6. The conventional method of expressing architectural measurement on a drawing is to show feet, inches, and fractions of an inch. True or False?

Answer: True

Question Type: T
Random: Y
Difficulty Level: 1
Chapter Number: 4
Page Reference: 74

7. Why are the basic structural features of a plan the first dimensions to be given consideration?

Answer: The basic structural features of a plan are the first dimensions given consideration because the structure is the basic organizing element of the building. Location of columns and load-bearing walls should be first priority.

Question Type: E
Random: Y
Difficulty Level: 2
Chapter Number: 4
Page Reference: 75

8. Why is it preferable to use only those abbreviations that have

been standardized and commonly acceptable?

Answer: Many people working in the construction industry need to read architectural drawings, so it is preferable to use only standard abbreviations which are recognized by the majority of those reading the drawings.

Question Type: E
Random: Y
Difficulty Level: 2
Chapter Number: 4
Page Reference: 78

9. What is the advantage of using a schedule when considerable information is involved?

Answer: A schedule offers the advantage of consolidating an extensive amount of similar information in a single location on drawings thus making the information easily accessible to estimators and builders.

Question Type: E
Random: Y
Difficulty Level: 2
Chapter Number: 4
Page Reference: 84

10. How can pencil smudging be removed from tracings?

Answer: Pencil smudges are difficult to remove from a pencil drawing. They can be erased with a soft eraser using an eraser shield. Care should be taken not to erase important line work.

Question Type: S
Random: Y
Difficulty Level: 2
Chapter Number: 4
Page Reference: 82

11. Describe three techniques used by drafters to keep drawings neat while drafting.

Answer: Using a dry cleaning pad, lifting drafting instruments over the drawing, and covering the finished part of a drawing with tracing paper are three techniques used by drafters to keep drawings neat while drafting.

Question Type: E

Random: Y
Difficulty Level: 2
Chapter Number: 4
Page Reference: 83-4

12. Which roof pitch is steeper, a 4/12 or a 8/12?

Answer: 8/12

Question Type: S
Random: Y
Difficulty Level: 2
Chapter Number: 4
Page Reference: 83

13. On floor plans of commercial buildings, dimensions are shown from wall to wall. How are the corresponding dimensions shown on residential floor plans?

Answer: Dimensions on residential floor plans are usually shown from the centerline of partition to the centerline of partition.

Question Type: S
Random: Y
Difficulty Level: 2
Chapter Number: 4
Page Reference: 77

14. Why is it necessary to draw a north arrow on floor plans?

Answer: The north arrow is shown on floor plans for orientation of the building.

Question Type: S
Random: Y
Difficulty Level: 2
Chapter Number: 4
Page Reference: 83

15. An arrowhead is used at the ends of dimension lines, what other symbol may be used at the ends of dimension lines?

Answer: A dot or tick can also be used at the intersection of dimension lines.

Question Type: S
Random: Y
Difficulty Level: 2

Chapter Number: 4
Page Reference: 75

16. Identify two types of schedules used on working drawings and describe how they are organized.

Answer: Room finish schedules used on working drawings are organized in vertical columns showing room numbers, room names, and finishes for walls, ceilings, and floors. Door and window schedules used on working drawings are organized in vertical columns showing door or window identification, types, and sizes.

Question Type: E
Random: Y
Difficulty Level: 2
Chapter Number: 4
Page Reference: 84

17. Why should the weight or width of linework vary on construction documents?

Answer: The significance of each line on the construction documents varies and is reflected in the weight or width of the line. If a line has little importance, it should appear light on the drawing. If the line contributes an important aspect to the drawing, it should appear heavy and prominent.

Question Type: E
Random: Y
Difficulty Level: 2
Chapter Number: 4
Page Reference: 71

18. To gain extreme emphasis, the ________________________ can be used on a few important lines on a drawing.

Answer: Technical drafting pen

Question Type: F
Random: Y
Difficulty Level: 1
Chapter Number: 4
Page Reference: 71

19. Architectural linework should end exactly at the corners. True

or False?

Answer: True

Question Type: T
Random: Y
Difficulty Level: 1
Chapter Number: 4
Page Reference: 77

20. What is an "extension" line?

Answer: An extension line is a light line drawn from the extremities of a feature requiring a dimension and should not touch the feature.

Question Type: S
Random: Y
Difficulty Level: 1
Chapter Number: 4
Page Reference: 74

21. What is a "dimension" line?

Answer: A dimension line is a light line that has arrowheads at each end indicating the exact extremities of the dimension.

Question Type: S
Random: Y
Difficulty Level: 1
Chapter Number: 4
Page Reference: 74

22. Where should the numerical dimension be placed in relation to the extension and dimension lines?

Answer: The numerical dimension should be placed approximately half way between the extension lines and slightly above the dimension line. The numerical dimension should never touch the dimension line.

Question Type: S
Random: Y
Difficulty Level: 1
Chapter Number: 4
Page Reference: 75

23. What is the conventional method of expressing architectural measurement with numbers?

Answer: The conventional method of expressing architectural measurement with numbers is feet, inches, and fractions of an inch (12' -4 1/2").

Question Type: S
Random: Y
Difficulty Level: 2
Chapter Number: 4
Page Reference: 74

24. Why is a series of dimensions "lined-up" on an architectural drawing?

Answer: A series of dimensions is lined up on an architectural drawing so they can be quickly drawn with one setting of the straight edge, so they will simplify the appearance or the drawing and so the overall total of the series of dimensions may be compared to the actual overall total dimension.

Question Type: S
Random: Y
Difficulty Level: 2
Chapter Number: 4
Page Reference: 75

25. Dimensions on architectural drawings are either ______________ or ____________________dimensions.

Answer: Size dimensions or location dimensions.

Question Type: F
Random: Y
Difficulty Level: 1
Chapter Number: 4
Page Reference: 76

26. What are "leaders"?

Answer: Leaders are straight or curved lines leading from a numerical dimension or note to the applying feature. An arrowhead or dot is usually drawn at the feature-end of the leader.

Question Type: S
Random: Y
Difficulty Level: 1
Chapter Number: 4
Page Reference: 76

27. What is the purpose for using angular or offset-curve leaders?

Answer: Angular or offset-curve leaders should appear as connectors only and should not be misinterpreted.

Question Type: S
Random: Y
Difficulty Level: 2
Chapter Number: 4
Page Reference: 76

28. For clarity of work required, notes should be placed on the drawings and in the written specifications. True or False?

Answer: False

Question Type: T
Random: Y
Difficulty Level: 3
Chapter Number: 4
Page Reference: 78

29. Why is it important to include a "legend" of unusual abbreviations on the architectural drawings?

Answer: It is important to include a legend of abbreviations on architectural drawings to reduce the possibility of a mix-understanding of the instructions caused by an unusual abbreviation.

Question Type: S
Random: Y
Difficulty Level: 3
Chapter Number: 4
Page Reference: 78

30. Discuss the purpose of material symbols.

Answer: Because most buildings are a complexity of different materials and components, a simplified method of representation is necessary on the drawings. A series of standard material symbols has been devised for the industry to alleviate this problem. If non-standard symbols are used, a symbols legend should be included. These non-standard symbols should be indicative of the material represented so not to be misleading.

Question Type: E
Random: Y
Difficulty Level: 3
Chapter Number: 4
Page Reference: 78

31. Define a "Schedule" on an architectural drawing.

Answer: A schedule is an organized arrangement of notes and/or information usually lettered within a lined enclosure.

Question Type: S
Random: Y
Difficulty Level: 1
Chapter Number: 4
Page Reference: 84

32. Discuss the purpose of a schedule?

Answer: The purpose of a schedule is to move extensive similar information to one location on the drawing instead of placing the information throughout the drawings. This makes the information easily available to the estimators and builders and reduces the potential of oversight by them.

Question Type: E
Random: Y
Difficulty Level: 2
Chapter Number: 4
Page Reference: 84

33. Windows are usually identified by __________, both in the schedule and near the window locations on the plan and elevations.

Answer: Letters.

Question Type: F
Random: Y
Difficulty Level: 2
Chapter Number: 4
Page Reference: 84

34. Doors are usually identified by the use of __________, both in the schedule and near their locations on the floor plans and elevations.

Answer: Numbers.

Question Type: F
Random: Y
Difficulty Level: 2
Chapter Number: 4
Page Reference: 84

35. Two (2) types of schedules used on residential working drawings are ______________________ and ______________________________.

Answer: Finish schedule and door or window schedules.

Question Type: F
Random: Y
Difficulty Level: 2
Chapter Number: 4
Page Reference: 84

36. Why are dimensions given to the center-line of door and window openings in frame construction?

Answer: Dimensions are given to the center-line of door and window openings on frame construction drawings because their exact location is not critical and the framing material is easily adaptable to fractional dimensions.

Question Type: S
Random: Y
Difficulty Level: 2
Chapter Number: 4
Page Reference: 76

37. Why are dimensions to door and window openings given to the rough opening of masonry construction?

Answer: Dimensions are given to the rough openings of masonry construction because of the material's uniform size, its permanence and difficulty in correcting mistakes, the coursing of the material and because it is usually the finished exterior material.

Question Type: S
Random: Y
Difficulty Level: 2
Chapter Number: 4
Page Reference: 76

38. Draw the symbol for a 5:12 roof slope.

Answer: 5/12 (see elevation drawing p.83)

Question Type: S
Random: Y
Difficulty Level: 2
Chapter Number: 4
Page Reference: 83

39. Why are architectural working drawings made on either tracing paper, vellum or film?

Answer: Because architectural working drawings are a means of communicating to many people, they have little value unless they can be satisfactorily and economically reproduced. Using a transparent medium allows duplication of the working drawings by different processes without destroying the original document.

Question Type: S
Random: Y
Difficulty Level: 2
Chapter Number: 4
Page Reference:

40. On a sectional drawing, what lineweigths should be emphasized?

Answer: Materials that have been cut through should be emphasized.

Question Type: S
Random: Y
Difficulty Level: 2
Chapter Number: 4
Page Reference: 73

41. For drawing difficult symbols the drafter may use a(n) ___________.

Answer: Template

Question Type: S
Random: Y
Difficulty Level: 2
Chapter Number: 4
Page Reference: 82

42. One should duplicate as many as possible dimensions on a drawing to save time for contractor. True or False?

Answer: False

Question Type: T
Random: Y
Difficulty Level: 2
Chapter Number: 4
Page Reference: 76

43. Sketch the symbol for concrete when in section.

Answer: (see p.79)

Question Type: S
Random: Y
Difficulty Level: 2
Chapter Number: 4
Page Reference: 79

44. Sketch the symbol for earth when in section.

Answer: (see p.79)

Question Type: S
Random: Y
Difficulty Level: 2
Chapter Number: 4
Page Reference: 79

45. Sketch the symbol for CMU's when in section.

Answer: (see p.79)

Question Type: S
Random: Y
Difficulty Level: 2
Chapter Number: 4
Page Reference: 79

46. Sketch the symbol for batt insulation when in section.

Answer: (see p.79)

Question Type: S
Random: Y
Difficulty Level: 2
Chapter Number: 4
Page Reference: 79

47. Sketch the symbol for finish wood when in section.

Answer: (see p.79)

Question Type: S
Random: Y
Difficulty Level: 2
Chapter Number: 4
Page Reference: 79

48. Sketch the symbol for rough wood when in section.

Answer: (see p.79)

Question Type: S
Random: Y
Difficulty Level: 2
Chapter Number: 4
Page Reference: 79

49. Sketch the symbol for field stone in elevation.

Answer: (see p.79)

Question Type: S
Random: Y
Difficulty Level: 2
Chapter Number: 4
Page Reference: 80

50. Dimensions should read from each of the four sides of a drawing. True or False?

Answer: False

Question Type: T
Random: Y

Difficulty Level: 1
Chapter Number: 4
Page Reference: 74

MULTIPLE CHOICE: Circle the letter indicating the best choice for each question.

51. Broken lines on architectural drawings can represent ___________.
    a. Hidden surfaces.
    b. Alternate positions of equipment.
    c. Structural pieces.
    d. Provide additional information.
    e. All of the above.

Answer: e

Question Type: F
Random: Y
Difficulty Level: 1
Chapter Number: 4
Page Reference: 75

Chapter 5. Modular and Metric Drafting

1. There are four aspects of modular coordination used in the building industry. They are :

    1. ____________________
    2. ____________________
    3. ____________________
    4. ____________________

Answer: 1. Think modular during all stages of the design process.
2. Draw the working drawings so that modular concepts are represented simple, but with flexibility.
3. Size building materials that conform with modular assembly.
4. Devise construction techniques that will utilize materials with minimal waste.

Question Type: F
Random: Y
Difficulty Level: 3
Chapter Number: 5
Page Reference: 91

2. The main unit of length in a metric system is ______________ .

Answer: 100 mm

Question Type: F
Random: Y
Difficulty Level: 2
Chapter Number: 5
Page Reference: 101

3. Define "soft" conversion in relation to using metrics in construction.

Answer: Where there is no dimensional change in the product, but the nearest practical metric dimension is used to express the metric equivalent.

Question Type: S
Random: Y
Difficulty Level: 2
Chapter Number: 5

Page Reference: 99

4. Define "hard" conversion in relation to using metrics in construction.

Answer: This implies making dimensional changes to a product.

Question Type: S
Random: Y
Difficulty Level: 2
Chapter Number: 5
Page Reference: 99

5. The basic module used in modular construction is a ______________ inch cube.

Answer: 4"

Question Type: F
Random: Y
Difficulty Level: 1
Chapter Number: 5
Page Reference: 92

6. Discuss modular coordination in vertical dimensioning.

Answer: Modular coordination should work with floor lines, ceiling lines, materials, and door & window heights.

Question Type: E
Random: Y
Difficulty Level: 3
Chapter Number: 5
Page Reference: 94

7. The basic module used in metric construction is ______________ .

Answer: 100 mm

Question Type: F
Random: Y
Difficulty Level: 1
Chapter Number: 5
Page Reference: 101

8. When using plywood as a dominant material, a logical grid size for preliminary design sketches would be ____________________ .

Answer: 4'-0" x 4'-0"

Question Type: F
Random: Y
Difficulty Level: 1
Chapter Number: 5
Page Reference: 91

9. When using concrete masonry units as a dominant material, a logical grid size for preliminary design sketches would be ____________________ .

Answer: 16"

Question Type: F
Random: Y
Difficulty Level: 2
Chapter Number: 5
Page Reference: 92

10. When dimensioning in the modular grid system, modular grid dimensions are indicated by ______________________ .

Answer: Arrowheads on both ends on dimension line.

Question Type: F
Random: Y
Difficulty Level: 2
Chapter Number: 5
Page Reference: 94

11. When dimensioning in the modular grid system, nongrid dimensions are indicated by ______________________ .

Answer: A dot on both ends on dimension line.

Question Type: F
Random: Y
Difficulty Level: 2
Chapter Number: 5
Page Reference: 94

12. When dimensioning in the modular grid system, nonmodular dimensions are indicated ______________________ .

Answer: With an arrowhead on one end and a dot on the other end of the dimension line.

Question Type: F
Random: Y
Difficulty Level: 2
Chapter Number: 5
Page Reference: 94

13. Metric units used for distances is ______________________ .

Answer: meters

Question Type: F
Random: Y
Difficulty Level: 1
Chapter Number: 5
Page Reference: 98

14. Metric units used for mass is ______________________ .

Answer: kilograms

Question Type: F
Random: Y
Difficulty Level: 2
Chapter Number: 5
Page Reference: 98

15. Metric units used for time is ______________________ .

Answer: seconds

Question Type: F
Random: Y
Difficulty Level: 1
Chapter Number: 5
Page Reference: 98

16. Metric unit used for electrical current is ________________.

Answer: ampere

Question Type: F
Random: Y
Difficulty Level: 2
Chapter Number: 5

Page Reference: 98

17. Metric unit used for temperature is ________________________ .
Answer: kelvin

Question Type: F
Random: Y
Difficulty Level: 2
Chapter Number: 5
Page Reference: 98

18. Metric unit used for illumination is ________________________ .
Answer: candela

Question Type: F
Random: Y
Difficulty Level: 2
Chapter Number: 5
Page Reference: 98

19. What is the purpose of reference grids?

Answer: Grids assist designers in the coordination of building dimensions.

Question Type: S
Random: Y
Difficulty Level: 1
Chapter Number: 5
Page Reference: 104

20. Define the Basic Modular Metric Grid.

Answer: Basic modular grids are those consecutive parallel lines spaced 100 mm apart horizontally and vertically.

Question Type: S
Random: Y
Difficulty Level: 1
Chapter Number: 5
Page Reference: 104

21. Define the Multi-modular Metric Grid.

Answer: Multi-modular grids are those used to layout a building and the location main structural elements.

Question Type: S
Random: Y
Difficulty Level: 2
Chapter Number: 5
Page Reference: 104

22. Define the Special Modular Metric Grid.

Answer: Special modular grids are tartan grids that alternate bands of 100 mm and 200 mm in plan.

Question Type: S
Random: Y
Difficulty Level: 2
Chapter Number: 5
Page Reference: 104

23. Define dimensional coordination.

Answer: Dimensional coordination is taking a basic module and applying a set of selected multiples or submultiples.

Question Type: S
Random: Y
Difficulty Level: 2
Chapter Number: 5
Page Reference: 101

24. International Systems, or SI, is commonly referred to as metric. True or False?

Answer: True

Question Type: T
Random: Y
Difficulty Level: 2
Chapter Number: 5
Page Reference: 97

25. Details of buildings do not use modular dimensions because of their uneven heights. True or False?

Answer: False

Question Type: T
Random: Y
Difficulty Level: 1
Chapter Number: 5
Page Reference: 93

26. The basic unit of measure in modular design is a one foot cube.
True or False?

Answer: False

Question Type: T
Random: Y
Difficulty Level: 1
Chapter Number: 5
Page Reference: 93

Chapter 6. BASIC TECHNICAL DRAWING

1. Describe the term orthographic projection.

Answer: An image created by lines that are projected from an object to a transparent plane out from the object is known as an orthographic projection. Orthographic projections commonly create the front, sides, and rear elevation views of an object; however, the top and bottom views are also considered orthographic views.

Question Type: E
Random: Y
Difficulty Level: 2
Chapter Number: 6
Page Reference: 107

2. What is a frontal plane?

Answer: The frontal plane is a vertical plane which is positioned out from the front elevation of an object. By the orthographic projection process the front of the object is drawn on the frontal plane.

Question Type: S
Random: Y
Difficulty Level: 2
Chapter Number: 6
Page Reference: 110

3. Why do drawings of many objects require more than one orthographic view?

Answer: Many objects are complicated in shape and form; therefore, more than one orthographic view is required to adequately describe the object, such as top, bottom, front, rear, and sides.

Question Type:
Random: Y
Difficulty Level:
Chapter Number: 6
Page Reference: 110

4. Which orthographic view is considered the major view?

Answer: The front elevation of an object is usually considered the major orthographic view.

Question Type: S
Random: Y
Difficulty Level: 1
Chapter Number: 6
Page Reference: 110

5. How are hidden surfaces shown on an orthographic view?

Answer: Hidden surfaces are shown as dashed (hidden) or broken lines on orthographic views.

Question Type: S
Random: Y
Difficulty Level: 1
Chapter Number: 6
Page Reference: 110

6. Why are orthographic views of a building often drawn in an unrelated position on working drawings?

Answer: Orthographic views of a building are often drawn in an unrelated position on working drawings so that all the elevation views would have the floor level parallel with the bottom of the sheet thus presenting the views as one would see them in reality.

Question Type: S
Random: Y
Difficulty Level: 2
Chapter Number: 6
Page Reference: 110

7. What is the cutting plane of a section?

Answer: The cutting plane is an imaginary plane that slices through an object identifying the place at which materials have been removed to reveal a sectional view of the remaining part of the object.

Question Type: S
Random: Y
Difficulty Level: 1
Chapter Number: 6
Page Reference: 115

8. Why are cutting planes sometimes offset rather than straight planes?

Answer: Cutting planes are sometimes offset to show special features about the object that cannot be revealed if the cutting plane passes through the object on a continuous straight plane.

Question Type: S
Random: Y
Difficulty Level: 2
Chapter Number: 6
Page Reference: 116

9. Why are hidden lines seldom used in section views?

Answer: Hidden or dashed lines are seldom shown on section views since these lines may become confusing to the viewer.

Question Type: S
Random: Y
Difficulty Level: 2
Chapter Number: 6
Page Reference: 110

10. What is the advantage of a partial section?

Answer: A partial has the advantage of permitting the viewer to see both the elevation and section view of an object on the same drawing or detail.

Question Type: S
Random: Y
Difficulty Level: 2
Chapter Number: 6
Page Reference: 117

11. Why are auxiliary surfaces projected to planes that are parallel to the surface?

Answer: Auxiliary surfaces are projected to planes parallel to the surfaces to prevent distortion of the surface when drawn.

Question Type: S
Random: Y
Difficulty Level: 1
Chapter Number: 6

Page Reference: 118

12. How are numerous sections identified as to their position on the major view?

Answer: Most sections on the major view will be vertical sections and are identified by a section cut symbol on the front elevation (major view).

Question Type: S
Random: Y
Difficulty Level: 2
Chapter Number: 6
Page Reference: 115

13. Describe a transverse section.

Answer: A vertical section, if completely taken through the narrower width of a building is called a transverse section.

Question Type: S
Random: Y
Difficulty Level: 2
Chapter Number: 6
Page Reference: 117

14. Describe a longitudinal section.

Answer: A vertical section, if completely taken through the long dimension of the building is called a longitudinal section.

Question Type: S
Random: Y
Difficulty Level: 2
Chapter Number: 6
Page Reference: 117

15. How many orthographic views can be made from a cube?

Answer: Six orthographic views can be made from a cube: top, bottom, front, rear, right, and left sides.

Question Type: S
Random: Y
Difficulty Level: 1
Chapter Number: 6
Page Reference: 109

16. What two views of a house could be revealed if a horizontal cutting plane passes midway between the floor and ceiling and completely bisects the house?

Answer: The floor plan and the reflected ceiling plan may be revealed if a horizontal cutting plane passes midway between the floor and ceiling and completely bisects the house.

Question Type: S
Random: Y
Difficulty Level: 2
Chapter Number: 6
Page Reference: 111

17. Why is it important to draw a vertical wall section from the footing to the roof of a house?

Answer: It is important to draw a small section from the footing to the roof of a house to show all the materials of construction in detail.

Question Type: S
Random: Y
Difficulty Level: 2
Chapter Number: 6
Page Reference:

18. Why are pictorial sections used?

Answer: Sometimes pictorial sections are used because they may be the most effective way to show structural assemblies or other construction features that are awkward to explain with orthographic views.

Question Type: S
Random: Y
Difficulty Level: 2
Chapter Number: 6
Page Reference: 108 & chapter 7 also.

19. What is a Golden Section?

Answer: The golden section. is a Greek method of dividing lines and forms into pleasing proportions. The ratio of one division to the other can be stated as 1 to 1.618.

Question Type: S

Random: Y
Difficulty Level: 2
Chapter Number: 6
Page Reference: 127

20. Orthographic projection serves as the conventional and universally accepted method for representing 3-dimensional objects on working drawings. True or False?

Answer: True

Question Type: T
Random: Y
Difficulty Level: 2
Chapter Number: 6
Page Reference: 111

21. Why are sections necessary on working drawings?

Answer: Sections are necessary to show information about construction that is ordinarily hidden on exterior views.

Question Type: S
Random: Y
Difficulty Level: 2
Chapter Number: 6
Page Reference: 111

22. What is the purpose of auxiliary projections?

Answer: Auxiliary projections show views that lie on other than orthographic planes.

Question Type: S
Random: Y
Difficulty Level: 2
Chapter Number: 6
Page Reference: 119

23. What is the "Golden Rectangle"?

Answer: The "Golden Rectangle" is a rectangle whose sides have a numerical proportion of 76:47 units. The short side of the rectangle may be laid out within the rectangle to form a square and the remaining space will always form another Golden Rectangle.

Question Type: S
Random: Y
Difficulty Level: 2
Chapter Number: 6
Page Reference: 127

24. In the "Glass Box" method of orthographic projection, what is the total number of views that may be developed?

Answer: The total number of views that may be developed from the "Glass Box" method of orthographic projection is six (6).

Question Type: S
Random: Y
Difficulty Level: 1
Chapter Number: 6
Page Reference: 110

25. In the "Glass Box" method of orthographic projection how many views are required to be known in order to develop the remaining views?

Answer: The minimum number of views required to be known in order to develop the remaining views are two (2).

Question Type: S
Random: Y
Difficulty Level: 2
Chapter Number: 6
Page Reference: 110

26. In orthographic projection, the most top view is considered to be the most important. True or False?

Answer: False

Question Type: T
Random: Y
Difficulty Level: 2
Chapter Number: 6
Page Reference: 110

27. In the "Glass Box" method of orthographic projection which three

(3) views may be considered as unnecessary?

Answer: The three views that may be considered as unnecessary in orthographic projection are the bottom view, the left side view and the rear view.

Question Type: S
Random: Y
Difficulty Level: 2
Chapter Number: 6
Page Reference: 110

28. Describe the third angle method of orthographic projection.

Answer: In the third angle method of projection the three conventional views (front, top and right side) are placed within the third octant of the transparent plane. The top view is then projected to the horizontal plane, the front view is projected to the frontal plane and the right side view is projected to the profile plane.

Question Type: E
Random: Y
Difficulty Level: 2
Chapter Number: 6
Page Reference: 110

29. How are surfaces or features that are hidden from the observer represented in orthographic drawings?

Answer: Surfaces or features that are hidden from the observer are represented by dashed lines on orthographic drawings.

Question Type: S
Random: Y
Difficulty Level: 2
Chapter Number: 6
Page Reference: 110

30. The main drawing of a set of working drawings is the ___________.

Answer: Floor Plan

Question Type: F
Random: Y
Difficulty Level: 1
Chapter Number: 6

Page Reference: 111

31. What is the definition of a section?

Answer: A section is defined as an imaginary cut made through an object, or a combination of objects, with the material in front of the cut removed.

Question Type: S
Random: Y
Difficulty Level: 1
Chapter Number: 6
Page Reference: 114

32. Why are sections important in the transferral of information?

Answer: The section cut becomes a direct method of revealing important internal information about an object and therefore is important in the transferral of this information.

Question Type: S
Random: Y
Difficulty Level: 2
Chapter Number: 6
Page Reference: 114

33. Sections are usually cut through which of the following areas?

a. Floors
b. Walls
c. Doors
d. Stairs
e. Fireplaces
f. All of the above

Answer: f

Question Type: M
Random: Y
Difficulty Level: 2
Chapter Number: 6
Page Reference: 110

34. Discuss why dashed (hidden) lines are not indicated in architectural section cuts?

Answer: The dashed lines would confuse the overall section and therefore would be useless information. Additional section cuts are made for those hidden areas requiring more explanation.

Question Type: E
Random: Y
Difficulty Level: 2
Chapter Number: 6
Page Reference: 110

35. How is the cutting-plane line usually designated on a set of working drawings?

Answer: The cutting-plane line is usually represented by a heavy dash and double-dot line at the location of the section. An arrow is usually placed at the end of the cutting-plane line to indicate the direction of observation revealed by the section view.

Question Type: E
Random: Y
Difficulty Level: 2
Chapter Number: 6
Page Reference: 115

36. What is crosshatching?

Answer: Crosshatching is the application of material symbols to sections designating the different materials and their relationship to one another.

Question Type: S
Random: Y
Difficulty Level: 2
Chapter Number: 6
Page Reference: 115

37. What does the word "poche" mean?

Answer: The word poche means to shade, color or otherwise darken the value of an area to provide contrast.

Question Type: S
Random: Y
Difficulty Level: 2

Chapter Number: 6
Page Reference: 116

38. On symmetrically shaped objects, half-sections conveniently show both the ______________________________ and the ______________________________ in only one view.

Answer: Interior and exterior

Question Type: F
Random: Y
Difficulty Level: 2
Chapter Number: 6
Page Reference: 117

39. Broken-out sections could be used to show important interior features that can be given sufficient description in a small area of an exterior view. Where could broken-out sections be utilized?

Answer: Broken-out sections could be used to show important interior features that can be given sufficient description in a small area of an exterior view.

Question Type: S
Random: Y
Difficulty Level: 2
Chapter Number: 6
Page Reference: 117

40. What is the difference between a revolved section and a removed section?

Answer: A revolved section is a partial section that has been rotated directly upon the object being cut and clearly shows the projection and contour of the elevation surface without the need for a separate view. A removed section is similar to the revolved section except it is not superimposed on the object but removed to a convenient location near the cutting plane.

Question Type: S
Random: Y
Difficulty Level: 2
Chapter Number: 6

Page Reference: 117

41. What is an auxiliary view?

Answer: An auxiliary view is a view projected to any plane other than one of the six orthographic planes and has been projected from a surface parallel to the auxiliary plane.

Question Type: S
Random: Y
Difficulty Level: 2
Chapter Number: 6
Page Reference: 119

42. Why are auxiliary views necessary?

Answer: Auxiliary views are necessary to reveal the true shape of an inclined surface.

Question Type: S
Random: Y
Difficulty Level: 2
Chapter Number: 6
Page Reference: 117

43. When are "true shape" views required?

Answer: "True shape" views are required when it is necessary to develop inclined oblique surfaces to determine the true length of members or areas.

Question Type: S
Random: Y
Difficulty Level: 2
Chapter Number: 6
Page Reference: 118

44. Why is it important that the drafter understand applied geometric construction, orthographic projection, sectioning and auxiliary projection?

Answer: It is important that the drafter understand applied geometric construction, orthographic projection, sectioning and auxiliary projection because each of these areas are of equal importance and knowledge of them is needed for a solid foundation in drafting.

Question Type: E
Random: Y
Difficulty Level: 2
Chapter Number: 6
Page Reference:

45. Discuss why top views or roof plans are not normally shown on residential working drawings.

Answer: Top views or roof plans are developed to show complex roof structures, locations of skylights, roof drains and other important information. These items are normally not a part of a simple residential design, but may be included in complex designs.

Question Type: E
Random: Y
Difficulty Level: 3
Chapter Number: 6
Page Reference: 111

46. The horizontal cut to indicate the floor plan is usually taken immediately below the bottom of the windows. True or False?

Answer: False

Question Type: T
Random: Y
Difficulty Level: 2
Chapter Number: 6
Page Reference: 112-3

47. Cutting-planes of all sections should be indicated on the working drawings. True or False?

Answer: False.

Question Type: T
Random: Y
Difficulty Level: 2
Chapter Number: 6
Page Reference: 115

48. What is the purpose of a "Reference plane" in the development of an auxiliary view?

Answer: The purpose of a "reference plane" in an auxiliary view is to facilitate the construction of the view by having a common point of reference on both the original object and the developed auxiliary view.

Question Type: S
Random: Y
Difficulty Level: 2
Chapter Number: 6
Page Reference: 120

49. List five (5) geometric constructions useful to the architectural drafter discussed in text.

1. ________________
2. ________________
3. ________________
4. ________________
5. ________________

Answer:
1. Bisect a line
2. Divide a line
3. Draw tangents
4. Draw reverse curves
5. Draw regular octagon
6. Draw regular pentagon
7. Draw regular hexagon
8. Draw regular polygon
9. Draw ellipse

Question Type: S
Random: Y
Difficulty Level: 2
Chapter Number: 6
Page Reference: 121

50. Define "regular" as it pertains to polygons.

Answer: A regular polygon has all angles equal.

Question Type: S
Random: Y
Difficulty Level: 2
Chapter Number: 6
Page Reference: 124

51. Discuss how to enlarge an irregular object using a grid.

Answer: First draw a small grid over the original object. Next draw a larger grid with the same number of squares. Number the grid lines and duplicate the details of each of the small squares to the corresponding large square.

Question Type: E
Random: Y
Difficulty Level: 3
Chapter Number: 6
Page Reference: 126

52. Thin members, when in section, are usually filled-in solid black. True or False?

Answer: True

Question Type: T
Random: Y
Difficulty Level: 2
Chapter Number: 6
Page Reference: 116

53. Items such as wood screws and steel bolts are hatched with steel symbols. True or False?

Answer: False

Question Type: T
Random: Y
Difficulty Level: 2
Chapter Number: 6
Page Reference: 116

MULTIPLE CHOICE: Circle the letter indicating the best choice for each question.

54. The word "Ortho" is defined as:
  a. Vertical
  b. Perpendicular
  c. Horizontal
  d. Diagonal
  e. All of the above

Answer: b

Question Type: M

Random: Y
Difficulty Level: 2
Chapter Number: 6
Page Reference: 107

55. In the "Glass Box" method of orthographic projection which of the following statements is incorrect?
  a. The front view is directly over the top view
  b. The right side view is directly to the right of the front view.
  c. The left side view is directly to the right of the rear view.
  d. The bottom view is directly below the front view.
  e. None of the above

Answer: a

Question Type: M
Random: Y
Difficulty Level: 2
Chapter Number: 6
Page Reference: 109

56. Orthographic views that are projected to vertical planes are called ______________________________.
  a. Plans
  b. Details
  d. Elevations
  e. None of the above

Answer: d

Question Type: M
Random: Y
Difficulty Level: 2
Chapter Number: 6
Page Reference: 111

57. Orthographic views that are projected to horizontal planes are known as ______________________________.
  a. Plans
  b. Sections
  c. Details
  d. Elevations
  e. None of the above

Answer: a.

Question Type: M
Random: Y
Difficulty Level: 2
Chapter Number: 6
Page Reference: 110

58. A transverse section is a cut taken completely through an object in which of the following directions?
    a. Horizontally
    b. Vertically
    c. Through the narrower width of the object
    d. Through the length of the object
    e. Diagonally

Answer: c

Question Type: M
Random: Y
Difficulty Level: 2
Chapter Number: 6
Page Reference: 117

Chapter 7. AXONOMETRIC AND OBLIQUE PICTORIAL DRAWINGS

1. What is the advantage of a mechanical pictorial drawing over an orthographic drawing?

Answer: One mechanical pictorial drawing can explain An object in a three dimensional form much better than a series of two dimensional orthographic drawings.

Question Type: S
Random: Y
Difficulty Level: 2
Chapter Number: 7
Page Reference: 131

2. What are the applications of axonometric and oblique drawings on working drawings?

Answer: Axonometric and oblique drawings are easy to construct with instruments and are therefore used to represent some construction details in a three dimensional way on working drawings.

Question Type: S
Random: Y
Difficulty Level: 2
Chapter Number: 7
Page Reference: 131

3. Why does the oblique type of drawing have more versatility than the axonometric?

Answer: Oblique drawings have more versatility then Econometric drawings because one side or plan is shown in orthographic projection and an infinite number of angles may be used to show the receding fixes which define the top and side.

Question Type: S
Random: Y
Difficulty Level: 2
Chapter Number: 7
Page Reference: 138

4. What angle from the horizontal gives emphasis to the top view in an oblique drawing?

Answer: The 60 degree angle from the horizontal would place emphasis on the top of an object drawn in oblique.

Question Type: S
Random: Y
Difficulty Level: 1
Chapter Number: 7
Page Reference: 138

5. Describe a cabinet drawing.

Answer: A cabinet drawing is an oblique drawing which has a ratio of 1 to 1/2 and is generally too thin to show most objects other than shallow cabinets.

Question Type: S
Random: Y
Difficulty Level: 2
Chapter Number: 7
Page Reference: 138

6. Why is it advantageous to place the irregular shapes of an object on the frontal plane when making an oblique drawing?

Answer: It is advantageous to place the irregular shapes of an object on the frontal plane when making an oblique drawing because objects like circles can be drawn with a compass on the frontal plane in stead of an ellipse which is drawn on the oblique plane.

Question Type: E
Random: Y
Difficulty Level: 2
Chapter Number: 7
Page Reference: 139

7. How may a plan view of a building be quickly made into a pictorial drawing?

Answer: A plan drawing can be made into a pictorial drawing by using a "plan oblique" or "elevation oblique" method of pictorial drawing.

Question Type: S
Random: Y
Difficulty Level: 2
Chapter Number: 7
Page Reference: 143

8. In dimensioning pictorial drawings, why are leaders drawn at angles other than those of the principle axes?

Answer: In dimensioning pictorial drawings leaders should be drawn at angles to the principle axes so that they are not confused with dimension lines.

Question Type: S
Random: Y
Difficulty Level: 2
Chapter Number: 7
Page Reference: 144

9. How are dimension numbers made to appear as though they lie on the planes of the drawing?

Answer: Dimension numbers are made to appear as though they lie on the planes of the drawing by using vertical lettering, yet the vertical axis of the lettering is kept parallel with the extension lines of the dimensions.

Question Type: S
Random: Y
Difficulty Level: 2
Chapter Number: 7
Page Reference: 144

10. Are the longest dimensions placed closest to the view or farther from the view, as compared to the shorter dimensions?

Answer: The longest dimensions are placed farthest from the view.

Question Type: S
Random: Y
Difficulty Level: 1
Chapter Number: 7
Page Reference: 132

11. Dicuss the difference between the axonometric drawing and a perspective drawing?

Answer: An axonometric drawing is an accurately scaled pictorial drawing which has been rotated on its axis to give it a three dimensional appearance and appears distorted. A perspective is a three dimensional drawing which is developed with a horizon line, station point and vanishing point or points and results in a pictorial view which is normal to the eye and not distorted.

Question Type: E
Random: Y

Difficulty Level: 2
Chapter Number: 7
Page Reference:

12. What is the proper title of a pictorial drawing that has x, y, and z axes and two of the angles between adjacent axes are equal?

Answer: The proper title of a pictorial drawing that has x, y, and z axes and two of the angles between the adjacent axes are equal is a "dimetric" drawing.

Question Type: S
Random: Y
Difficulty Level: 2
Chapter Number: 7
Page Reference: 134

13. What view of an object is revealed by a reverse axes drawing?

Answer: The underside or bottom of an object is revealed by a reverse axes drawing.

Question Type: S
Random: Y
Difficulty Level: 1
Chapter Number: 7
Page Reference: 138

14. List the types of pictorial drawings that require a true-shape plan view that can be traced as a basis for the drawing?

Answer: Plan-oblique, vertical projection; plan oblique, angular projection; and planometric are drawings which require a true-shape plan view that can be traced as a basis for the drawing.

Question Type: S
Random: Y
Difficulty Level: 2
Chapter Number: 7
Page Reference: 141-3

15. Why are 30°, 45°, and 60° angles used in preparing most pictorial drawings?

Answer: The 30°, 45°, and 60° angles are used in preparing most pictorial drawings because these angles are standard angles of drafting triangles.

Question Type: S
Random: Y
Difficulty Level: 2
Chapter Number: 7
Page Reference: 139

16. What is the proper title of the pictorial drawing that has equal angles between the x, y, and z axes?

Answer: The "isometric" is a pictorial drawing that has equal angles between the x, y, and z axes.

Question Type: S
Random: Y
Difficulty Level: 2
Chapter Number: 7
Page Reference: 133

17. Define paraline drawings.

Answer: Paraline drawings are pictorial drawings that enable the drafter to create single view images which represent three dimensional space. Axonometric and oblique drawings are paraline drawing. The sides of the object on the drawing are represented by parallel lines, conveniently executed with the parallel bar and triangles.

Question Type: S
Random: Y
Difficulty Level: 2
Chapter Number: 7
Page Reference: 131

18. List the type of pictorial drawings that require true measurements for all lines making up the drawing.

Answer: Axonometric, isometric, diametric, trimetric, and planometric), and oblique (elevation-oblique and plan-oblique) are pictorial drawings that require true measurements for all lines making up the drawing.

Question Type: S
Random: Y
Difficulty Level: 3
Chapter Number: 7
Page Reference:

19. What is the difference between a plan oblique with vertical projection and a plan oblique with angular projection?

Answer: The plan oblique-vertical projection is executed by rotating a true-shape plan 45° or 30°-60° to the horizontal and using a triangle to draw all vertical lines perpendicular to the horizontal; where as, a plan oblique-angular projection is executed by placing a true-shape plan parallel to the picture plane and the vertical line. are projected upward at a 45° or 60° angle to the horizontal.

Question Type: S
Random: Y
Difficulty Level: 2
Chapter Number: 7
Page Reference: 141

20. What characteristic is lost in the planometric drawing that is evident in the plan-oblique drawing?

Answer: The planometric negates all right-angled side planes that are evident in the plan oblique drawing.

Question Type: S
Random: Y
Difficulty Level: 2
Chapter Number: 7
Page Reference: 131

21. Discuss the difference between orthographic drawings and Axonometric and oblique drawings?

Answer: Axonometric and oblique drawings are pictorial in nature, show three sides of the object and the principal lines are measured directly. Orthographic drawings are flat projections of a surface and only indicate the one surface.

Question Type: E
Random: Y
Difficulty Level: 2
Chapter Number: 7
Page Reference: 131

22. What are the advantages of Axonometric and oblique drawings?

Answer: Axonometric and oblique drawings are easy to freehand sketch or draw with instruments, are suitable for use on working drawings, can be occasional design drawings and are easily understood by most parties.

Question Type: S
Random: Y
Difficulty Level: 2
Chapter Number: 7
Page Reference: 131

23. What is an Axonometric drawing?

Answer: An Axonometric drawing is an accurately scaled drawing which depicts an object that has been rotated on its axis and inclined from the regular parallel position to give it a three dimensional appearance.

Question Type: S
Random: Y
Difficulty Level: 2
Chapter Number: 7
Page Reference: 132

24. The three (3) systems of Axonometric drawings that are commonly used are:

1. __________

2. __________

3. __________

Answer: 1. Isometric 2. Dimetric 3. Trimetric

Question Type: F
Random: Y
Difficulty Level: 2
Chapter Number: 7
Page Reference: 133-5

25. A characteristic of Axonometric axes is that neither the angle formed between the X and Y axes nor the angle formed between the Y and Z axes may be drawn at 90 degrees. True or False?

Answer: True

Question Type: T

Random: Y
Difficulty Level: 2
Chapter Number: 7
Page Reference: 132

26. Define an isometric drawing.

Answer: An isometric drawing is developed when the rotation and inclination of the planes results in a drawing with three equally divided angles about a center point.

Question Type:
Random: Y S
Difficulty Level: 2
Chapter Number: 7
Page Reference: 133

27. What are the four (4) basic rules to construct an axonometric drawing?

1. ____________________

2. ____________________

3. ____________________

4. ____________________

Answer: 1. All vertical lines are vertical 2. All horizontal lines are on an angle 3. All lines are parallel 4. All dimensions are true length

Question Type: F
Random: Y
Difficulty Level: 2
Chapter Number: 7
Page Reference: 132

28. Define a Dimetric drawing.

Answer: A Dimetric drawing has two equally divided angles which are not 90 degrees or 120 degrees at the intersection of the X, Y and Z axes.

Question Type: S
Random: Y
Difficulty Level: 2
Chapter Number: 7

Page Reference: 134

29. Define a Trimetric drawing.

Answer: A Trimetric drawing has three unequal angles intersecting at the X, Y and Z axes.

Question Type: S
Random: Y
Difficulty Level: 2
Chapter Number: 7
Page Reference: 135

30. Between the Isometric, Dimetric and Trimetric drawings, which has the most flexibility?

Answer: The Trimetric drawing has the most flexibility since greater emphasis may be placed on any of the planes. Whereas the dimetric drawing will emphasize two of the planes and the Isometric gives prominence to all three sides equally.

Question Type: E
Random: Y
Difficulty Level: 3
Chapter Number: 7
Page Reference: 133-5

31. When would a drafter employ a "reverse isometric"?

Answer: A drafter would employ a "Reverse Isometric" when it is necessary to show the underside of an object.

Question Type: S
Random: Y
Difficulty Level: 2
Chapter Number: 7
Page Reference: 138

32. Angular shapes appear as their exact shape on a three-dimensional drawing. True or False?

Answer: False

Question Type: T
Random: Y
Difficulty Level: 2

Chapter Number: 7
Page Reference: 135

33. An oblique drawing shows an object with two planes in true orthographic projection and one side receding at various angles. True or False?

Answer: False

Question Type: T
Random: Y
Difficulty Level: 1
Chapter Number: 7
Page Reference: 138

34. One of the disadvantages of oblique drawings is their apparent versatility in creating pictorial effects by varying the angle of the receding axis. True or False?

Answer: False

Question Type: T
Random: Y
Difficulty Level: 2
Chapter Number: 7
Page Reference: 139

35. The two (2) commonly used methods of oblique drawings are

_________________________ and ______________________________ .

Answer: Cavalier and Cabinet

Question Type: F
Random: Y
Difficulty Level: 2
Chapter Number: 7
Page Reference: 139

36. When the frontal view is orthographic, the sides recede at a 45 degree angle and all lines are true length the drawing is known as _________________________ a(n) drawing.

Answer: Cavalier oblique

Question Type: F
Random: Y

Difficulty Level: 2
Chapter Number: 7
Page Reference: 139

37. When the frontal view is orthographic, the sides recede at a 45 degree angle and the lines are scaled at 1/2 their true length the drawing is known as a(n)__________________________ drawing.

Answer: Cabinet oblique

Question Type: F
Random: Y
Difficulty Level: 2
Chapter Number: 7
Page Reference: 139

38. Why is the depth of a cabinet oblique drawing reduced?

Answer: The depth of a cabinet oblique drawing is reduced to have more pleasing proportions and a more realistic appearance on objects with great depth.

Question Type: S
Random: Y
Difficulty Level: 2
Chapter Number: 7
Page Reference: 139

39. Why should circles be placed on the frontal view in oblique drawings?

Answer: Circles may be drawn with a compass and do not have to be distorted when shown on the frontal view of an oblique drawing.

Question Type: S
Random: Y
Difficulty Level: 2
Chapter Number: 7
Page Reference: 139

40. If circles become necessary on the oblique axis, how should they be drawn?

Answer: If circles are necessary on the oblique axis a rhombus should be constructed around the intended circle and the

4-center method of ellipse construction used.

Question Type: S
Random: Y
Difficulty Level: 2
Chapter Number: 7
Page Reference: 137

41. What is the purpose of the plan oblique?

Answer: The purpose of the plan oblique is to give a more pictorial nature to a two-dimensional drawing and to indicate heights of various items.

Question Type: S
Random: Y
Difficulty Level: 2
Chapter Number: 7
Page Reference: 143

42. What are the two (2) sub-classes of the plan oblique?

Answer: The two sub-classes of the plan oblique are (1) plan oblique - vertical projection and (2) plan oblique - angular projection.

Question Type: S
Random: Y
Difficulty Level: 2
Chapter Number: 7
Page Reference: 141-143

43. Describe the plan oblique - vertical projection method of drawing.

Answer: When drawing the plan oblique - vertical projection method the orthographic floor plan is tilted at a 45 degree, a 30 degree or a 60 degree angle and the vertical lines or planes are drawn perpendicular to the horizontal.

Question Type: E
Random: Y
Difficulty Level: 2
Chapter Number: 7
Page Reference: 141

44. Describe the plan oblique - angular projection method of drawing.

Answer: In the plan oblique - angular projection method the plan is placed parallel to the picture plane and the vertical lines and planes are projected upward at an angle, usually 45 degrees or 60 degrees.

Question Type: E
Random: Y
Difficulty Level: 2
Chapter Number: 7
Page Reference: 143

45. Describe the planometric method of pictorial drawing.

Answer: In the planometric method of pictorial drawing the true-shaped plan is positioned with one of its sides parallel to the picture plane and the receding lines and planes are drawn perpendicular to the picture plane.

Question Type: E
Random: Y
Difficulty Level: 2
Chapter Number: 7
Page Reference: 144

46. When constructing paraline drawings the longest dimensions are placed closest to the view. True or False?

Answer: False

Question Type: T
Random: Y
Difficulty Level: 2
Chapter Number: 7
Page Reference: 144

47. In axonometric and oblique drawings visible surfaces and features should only be dimensioned. True or False?

Answer: True

Question Type: T
Random: Y
Difficulty Level: 2
Chapter Number: 7
Page Reference: 144

48. In axonometric and oblique drawings dimension lines may connect extension lines that lie on different surfaces. True or False?

Answer: False

Question Type: T
Random: Y
Difficulty Level: 2
Chapter Number: 7
Page Reference: 144

49. Isometric pictorial sections may be partially broken. True or False?

Answer: True

Question Type: T
Random: Y
Difficulty Level: 1
Chapter Number: 7
Page Reference: 138

50. Isometric pictorial sections tend to be confusing to the observer. True or False?

Answer: False

Question Type: T
Random: Y
Difficulty Level: 1
Chapter Number: 7
Page Reference: 138

MULTIPLE CHOICE: Circle the letter indicating the best choice for each question.

51. A true circle on a three-dimensional drawing becomes a(n) ________________ .

a. Three point arc
b. Ellipse
c. Circle
d. Polygon
e. None of the above

Answer: b

Question Type: M
Random: Y
Difficulty Level: 2
Chapter Number: 7
Page Reference: 137

52. When an important elevation of an object requires emphasis, which of the following methods should be utilized?
    a. Isometric
    b. Plan oblique
    c. Elevation oblique
    d. None of the above

Answer: c

Question Type: M
Random: Y
Difficulty Level: 2
Chapter Number: 7
Page Reference: 140

53. When dimensioning axonometric and oblique drawings which of the following should not be done?
    a. Leaders should be made parallel with the axes of the drawing.
    b. Leaders that dimension circular shapes should just touch the arc.
    c. Notes may be lettered on either of the pictorial planes.
    d. Extension lines and dimension lines should be kept on the surface in question.
    e. Dimension numbers should be kept just above the dimension line.

Answer: a

Question Type: M
Random: Y
Difficulty Level: 2
Chapter Number: 7
Page Reference: 144

Chapter 8. FREEHAND SKETCHING

1. Why does ability in sketching require a strong sense of observation?

Answer: Sketching ability requires a strong sense of observation, because the eye must be trained to see the correct proportion of how one part relates to the other.

Question Type: S
Random: Y
Difficulty Level: 2
Chapter Number: 8
Page Reference: 151

2. Explain the structural or skeleton method of developing a sketch.

Answer: Many objects have a basic structural system which should be studied before sketching the object. An example would be the need to study the human skeletal structure before drawing human body figures.

Question Type: E
Random: Y
Difficulty Level: 2
Chapter Number: 8
Page Reference: 157

3. Why is sketching a subject reduced to its simplest geometric forms in organizing the sketch and the details drawn later?

Answer: A sketching subject is reduced to its simplest geometric form in order to distinguish its proportion to scale.

Question Type: S
Random: Y
Difficulty Level: 2
Chapter Number: 8
Page Reference: 155

4. What type of practice sketches should precede pictorial sketches?

Answer: Pictorial sketches should be preceded by practice sketches of line-, angles, circle-, and arcs.

Question Type: S
Random: Y

Difficulty Level: 2
Chapter Number: 8
Page Reference: 152

5. Which simple geometric form aids in establishing proportion?

Answer: The simple geometric form of the square aids in establishing proportion.

Question Type: S
Random: Y
Difficulty Level: 1
Chapter Number: 8
Page Reference: 155

6. Why do architectural drafters prefer to do their sketching on inexpensive tracing paper?

Answer: Architectural drafters prefer to do their sketching on inexpensive tracing paper because of the ease with which sketches can be modified or altered simply by placing the transparent paper over previous sketches. Tracing is quicker than reconstructing or erasing. Numerous sketcher are needed before a final detail is created.

Question Type: E
Random: Y
Difficulty Level: 1
Chapter Number: 8
Page Reference: 151

7. For what reason is sketching a necessary part of drafting training?

Answer: Sketching is a necessary part of drafting training in order to generate ideas and to make concepts workable realities.

Question Type: S
Random: Y
Difficulty Level: 1
Chapter Number: 8
Page Reference:

8. Another name for a schematic sketch floor plan is ___________.

Answer: Bubble sketch.

Question Type: F
Random: Y
Difficulty Level: 2
Chapter Number: 8
Page Reference: 162

9. Why is the use of coordinate paper underlays especially helpful in sketching floor plans?

Answer: Coordinate paper underlays are helpful in sketching floor plans because they give a reference to scale.

Question Type: S
Random: Y
Difficulty Level: 2
Chapter Number: 8
Page Reference: 159

10. Why is it essential for the beginner to practice sketching basic geometric shapes?

Answer: It is essential for the beginner to practice sketching basic geometric shapes because they are the fundamental forms of architecture.

Question Type: S
Random: Y
Difficulty Level: 2
Chapter Number: 8
Page Reference: 154

11. Describe the method used for drawing a freehand circle using two pencils.

Answer: A freehand circle may be drawn by holding two pencils of the same length rigidly in the hand so that they act much as a compass. One pencil becomes the pivot; the other scribes the circle when the paper is revolved under the pencils.

Question Type: E
Random: Y
Difficulty Level: 2
Chapter Number: 8
Page Reference: 155

12. Why are shades and shadows essential elements of the sketch?

Answer: Shades and shadows are essential elements of the sketch to show realism.

Question Type: S
Random: Y
Difficulty Level: 2
Chapter Number: 8
Page Reference: 164

13. What advantage does the felt-tip pen sketch have over the pencil sketch when reproduction of the sketch is important?

Answer: The reproduction of the felt-tip pen sketch is sharper and darker than the pencil sketch.

Question Type: S
Random: Y
Difficulty Level: 2
Chapter Number: 8
Page Reference: 152

14. How is good line technique acquired when sketching?

Answer: Good line technique is developed with practice; each exercise should be a deliberate effort to improve.

Question Type: S
Random: Y
Difficulty Level: 2
Chapter Number: 8
Page Reference: 156

15. What method is recommended for sketching symmetrical objects?

Answer: To sketch a symmetrical object draw a centerline on the paper and proportion the object equally on either side of the centerline.

Question Type: S
Random: Y
Difficulty Level: 2
Chapter Number: 8
Page Reference: 154

16. Why is it important that drafters have freehand sketching abilities?

Answer: The drafter needs the ability to sketch so that they can communicate ideas quickly. Most detailed drawings are developed from sketches.

Question Type: S
Random: Y
Difficulty Level: 2
Chapter Number: 8

17. State four (4) reasons why a drafter should have sketching ability.

1. ______________________________

2. ______________________________

3. ______________________________

4. ______________________________

Answer: Sketching is:

1. A means of technical expression in generating ideas and making concepts workable.
2. An important exploratory instrument.
3. A method to analyze graphic problems.
4. Necessary in planning and organizing the sheet layout for a set of working drawings.

Question Type: F
Random: Y
Difficulty Level: 2
Chapter Number: 8

18. What grade of graphite-clay lead is recommended for freehand sketching?

Answer: A soft lead, usually an "H" or "HB" grade, is recommended for freehand sketching.

Question Type: S
Random: Y
Difficulty Level: 1
Chapter Number: 8
Page Reference: 151

19. The most useful type of paper for freehand sketching is

_____________________ or _____________________ .

Answer: Inexpensive tracing paper or a sketch pad.

Question Type: F
Random: Y
Difficulty Level: 1
Chapter Number: 8
Page Reference: 151

20. Light lines are obtained by using a harder pencil lead. True or False?

Answer: True

Question Type: T
Random: Y
Difficulty Level: 1
Chapter Number: 8
Page Reference: 151

21. Name two (2) methods of dividing a line into equal parts.

1. ____________________________________

2. ____________________________________

Answer: 1. Using a pencil, lay off approximate half-lengths of the line from each end and bisect the remaining spaces by eye.
2. By bisecting the line and subsequent spaces by visual comparison.

Question Type: F
Random: Y
Difficulty Level: 1
Chapter Number: 8
Page Reference: 153

22. How is the center of a square or rectangle found?

Answer: The center of a square or rectangle is found by drawing the diagonal lines from corner to opposite corner. The intersection of the two diagonals will be the center of the space.

Question Type: S
Random: Y
Difficulty Level: 2
Chapter Number: 8
Page Reference: 154

23. Describe how larger circles are sketched.

Answer: Larger circles are constructed by sketching a square whose sides are equal to the diameter of the required circle and dividing the square into four (4) smaller equal squares. From each end of each side of the larger square find a point that is 2/7 from each corner of each side. Connect these points with a diagonal (45 degree) line. Inside this approximate octagon sketch the circle.

Question Type: S
Random: Y
Difficulty Level: 2
Chapter Number: 8
Page Reference: 155

24. What is meant by a "schematic floor plan sketch"?

Answer: A schematic floor plan sketch is a drawing of a pre-preliminary design indicating relative sizes of various areas and their relationship to each other. The shapes of these areas may be any shape the designer chooses and details of each area are not included.

Question Type: S
Random: Y
Difficulty Level: 3
Chapter Number: 8
Page Reference: 162

25. Before sketching the exterior elevations of a building, it is helpful to develop the floor plan complete with roof shapes and overhangs. True or False?

Answer: True

Question Type: T
Random: Y
Difficulty Level: 2
Chapter Number: 8
Page Reference:

26. Why would the use of non-reproducible grid type tracing paper be helpful in sketching?

Answer: The use of non-reproducible grid type tracing paper is helpful in sketching because it (1) provides vertical and horizontal guide lines, (2) allows a scale to be easily maintained, (3) provides guidelines for lettering and does not allow the grid to show on the duplicated product.

Question Type: S
Random: Y
Difficulty Level: 2
Chapter Number: 8
Page Reference: 159

27. How is good line technique acquired in sketching?

Answer: Good line technique is acquired in sketching by constant practice. To gain proficiency situations entailing sketching must be invited at every opportunity.

Question Type: S
Random: Y
Difficulty Level: 1
Chapter Number: 8
Page Reference: 156

MULTIPLE CHOICE: Circle the letter indicating the best choice for each question.

28. What geometric shape makes a suitable unit for simplifying a subject to be sketched?
    a. Arc
    b. Circle
    c. Rectangle
    d. Square
    e. Triangle

Answer: d

Question Type: M
Random: Y
Difficulty Level: 2
Chapter Number: 8
Page Reference: 155

Chapters 9. PERSPECTIVE DRAWINGS

1. What is meant by linear perspective?

Answer: Linear perspectives are those defining outlines used by delineators and classified as either one-point, two-point, or three-point perspectives.

Question Type: S
Random: Y
Difficulty Level: 2
Chapter Number: 9
Page Reference: 175

2. How does the projection of a perspective view of the picture plane differ from the projection of an orthographic view?

Answer: The basic theory in perspective drawing assumes that the image is produced on a transparent vertical plane called the picture plane very much as orthographic views are formed on transparent planes. The only difference is the orthographies have parallel projectors, whereas the perspective views have projectors radiating from a single person's eye.

Question Type: E
Random: Y
Difficulty Level: 2
Chapter Number: 9
Page Reference: 170

3. What is the point called that approximates the eye when a person draws an object in perspective?

Answer: The station point is the point that approximates the eye when a person draws an object in perspective.

Question Type: S
Random: Y
Difficulty Level: 2
Chapter Number: 9
Page Reference: 170

4. What effect is produced on a perspective when the horizon line is raised in relation to the ground line?

Answer: The birds eye or "top" view is produced on a perspective when the horizon is raised in relation to the groundline.

Question Type: S
Random: Y
Difficulty Level: 1
Chapter Number: 9
Page Reference: 175

5. What effect is produced on a perspective when the object is placed farther away from the picture plane?

Answer: Objects appear smaller the farther back they are from the picture plane.

Question Type: S
Random: Y
Difficulty Level: 1
Chapter Number: 9
Page Reference: 173

6. Why must true heights of an object be established only on the picture plane?

Answer: True heights of an object must be established only at the picture plane, since in perspective drawing, objects beyond the picture plane are smaller than true height and objects in front of it are larger than true height.

Question Type: E
Random: Y
Difficulty Level: 2
Chapter Number: 9
Page Reference: 173

7. If an object is placed in an oblique position in relation to the picture plane, why are two vanishing points necessary?

Answer: When a building is placed in an oblique position on a picture plane, the viewer would see two sides of the building; therefore there must be two vanishing points.

Question Type: E
Random: Y
Difficulty Level: 2
Chapter Number: 9
Page Reference: 174

8. What are the advantages of using the perspective plan method for

drawing a perspective?

Answer: The perspective-plan method has more versatility than the office-method. Drawings can be completed more quickly once the principles have been mastered since measurements can be placed on a measuring line eliminating the need to draw the floor plan above the perspective view and make projections from it.

Question Type: E
Random: Y
Difficulty Level: 2
Chapter Number: 9
Page Reference: 183

9. In a one-point parallel perspective view of an interior, is the single vanishing point used to draw a piece of furniture that is in a diagonal position? Explain.

Answer: Yes, there is one vanishing point in a one point parallel perspective view.

Question Type: E
Random: Y
Difficulty Level: 2
Chapter Number: 9
Page Reference: 188

10. Why are most two-point perspectives drawn with the horizon line located about 6' above the ground plane?

Answer: Most two-point perspectives are drawn with the horizon line located about 6' above the ground plane; this distance represents the eye level height of the average person standing on level ground.

Question Type: S
Random: Y
Difficulty Level: 2
Chapter Number: 9
Page Reference: 183

11. Why is it important to limit the cone of vision from the station point to the picture plane to an angle of 30° to 45°?

Answer: It is important to limit the-cone of vision from the station point to the picture plane to an angle of 30. to 45. to prevent distortion of the perspective.

Question Type: S
Random: Y
Difficulty Level: 2
Chapter Number: 9
Page Reference: 173

12. When would it be appropriate to use a bird's eye view perspective?

Answer: A birds eye view perspective would be used for a roof-top or aerial view of a building.

Question Type: S
Random: Y
Difficulty Level: 1
Chapter Number: 9
Page Reference: 175

13. Describe a convenient method for drawing a circle in a two-point perspective?

Answer: Draw a square in perspective that has sides of the same dimensions as the diameter of the circle. Find the center points of each side of the square. Construct arcs tangent to center points on the sides of the square. The circle will appear as an ellipse in perspective.

Question Type: E
Random: Y
Difficulty Level: 2
Chapter Number: 9
Page Reference: 184

14. What is a perspective drawing?

Answer: A perspective drawing is a pictorial method of representing a building or object, very much as the lens of a camera records an image on film.

Question Type: S
Random: Y
Difficulty Level: 2
Chapter Number: 9
Page Reference: 169

15. Discuss the theory of perspective.

Answer: The basic theory of perspective drawing assumes that the image is produced upon a transparent vertical plane (called the picture plane), similar to an orthographic drawing except that the orthographic drawing has parallel projectors, whereas the perspective views have projectors radiating from a single point. The outline of these points as they pierce the "picture plan" forms the perspective drawing.

Question Type: E
Random: Y
Difficulty Level: 3
Chapter Number: 9
Page Reference: 169

16. Define the picture plane.

Answer: The picture plane is a transparent vertical plane on which the perspective is drawn.

Question Type: S
Random: Y
Difficulty Level: 2
Chapter Number: 9
Page Reference: 170

17. If an object touches the picture plane on the plan it will appear above and behind the ground line in the perspective. True or False?

Answer: False

Question Type: T
Random: Y
Difficulty Level: 2
Chapter Number: 9
Page Reference: 172

18. A general rule is to place the station point approximately ______ as far away from the picture plane as the length of the building being drawn.

Answer: Twice

Question Type: F
Random: Y
Difficulty Level: 1
Chapter Number: 9
Page Reference:

19. What are the purposes of vanishing points in a perspective?

Answer: The purpose of vanishing points in a perspective is to locate the spot where lines converge. These points are found on the horizon line for horizontal projections and the vertical trace through the vanishing points for sloping surfaces.

Question Type: S
Random: Y
Difficulty Level: 2
Chapter Number: 9
Page Reference: 172

20. Discuss the "true-height line" (T.H.L.) in perspective drawing.

Answer: If a vertical line of the object touches the picture plane in plan view the line will appear at the same scale on the finished perspective, thus providing a convenient method of projecting true heights. From the true height line the various heights of the object are either diminished or increased by projecting to or from the vanishing points.

Question Type: S
Random: Y
Difficulty Level: 2
Chapter Number: 9
Page Reference: 172

21. The three (3) variables in perspective drawing are the relationship between the:

1. ____________________

2. ____________________

3. ____________________

Answer: 1. Station point 2. Picture plane 3. Object

Question Type: F
Random: Y
Difficulty Level: 2
Chapter Number: 9
Page Reference: 170

22. The 30/60 degree angles are convenient for laying out the plan in relation to the picture plane. True or False?

Answer: True

Question Type: T
Random: Y
Difficulty Level: 1
Chapter Number: 9
Page Reference: 180

23. How can the size of the perspective be adjusted without changing the relationship of the station point to the picture plane?

Answer: The size of the perspective may be adjusted without changing the relationship of the station point to the picture plane by moving the plan either up or down thereby placing the front corner of the building in front of or behind the picture plane. It moves the true height line to either another point of the plan on the picture plane or a point projected from the object to the picture plane.

Question Type: S
Random: Y
Difficulty Level: 3
Chapter Number: 9
Page Reference: 174

24. What effect is produced on a perspective when the horizon line is raised in relation to the ground line?

Answer: The effect that is produced when the horizon line is raised in relation to the ground line is that the viewer is looking down on the object and the perspective becomes an aerial view.

Question Type: S
Random: Y
Difficulty Level: 2
Chapter Number: 9
Page Reference: 175

25. If an object is placed in a perpendicular position in relation to the picture plane why is only one vanishing point necessary?

Answer: Only one vanishing point is necessary when the object is placed in a perpendicular position in relation to the picture plane because only the lines extending parallel to the viewer's line of sight will converge to a vanishing point on the horizon. Other lines that are perpendicular to the viewer will be shown as orthographic projections at their relative distance from the picture plane.

Question Type: S
Random: Y
Difficulty Level: 3
Chapter Number: 9
Page Reference: 188

26. The horizon line is usually placed between __________________ and __________________ above the ground line.

Answer: Five feet, six inches (5'-6") and six feet (6'-0").

Question Type: F
Random: Y
Difficulty Level: 2
Chapter Number: 9
Page Reference: 183

27. Why is the two-point angular perspective the most popular type for showing the exteriors of buildings?

Answer: The two-point perspective is the most popular type for showing the exterior of buildings because two sides of the building can be seen and the angular nature of these sides reveals the important information without excessive distortion.

Question Type: S
Random: Y
Difficulty Level: 2
Chapter Number: 9
Page Reference:

28. The two methods of constructing perspectives are:

1. ______________________

2. ______________________

Answer: 1. The common or office method (direct projection)
2. The perspective-plan method

Question Type: F
Random: Y
Difficulty Level: 2
Chapter Number: 9
Page Reference: 175

29. What occurs when the vanishing point of a one point perspective of a room interior is placed outside the view?

Answer: When the vanishing point of a one point room interior perspective is placed outside the view a portion of the room is blocked by the walls closest to the observer. This develops an incomplete view of the room.

Question Type: S
Random: Y
Difficulty Level: 3
Chapter Number: 9
Page Reference: 188

30. In a one-point interior perspective, is there a single vanishing point used to draw a piece of furniture that is in a diagonal position? Explain.

Answer: No. In a one-point interior perspective a piece of furniture placed in an angular position will have two separate vanishing points since both sides of the furniture will be receding.

Question Type: S
Random: Y
Difficulty Level: 3
Chapter Number: 9
Page Reference: 192

31. When drawing an interior two-point angular perspective how far should the station point be from the picture plane?

Answer: When setting up an interior two-point angular perspective keep the station point about twice as far from the picture plane as the greatest width of the plan being drawn.

Question Type: S
Random: Y
Difficulty Level: 2
Chapter Number: 9
Page Reference:

32. When would it be appropriate to use a three point perspective?

Answer: Since vertical lines also vanish to a point in infinity, it would be appropriate to use a three point perspective when drawing tall buildings or objects.

Question Type: S
Random: Y
Difficulty Level: 3
Chapter Number: 9
Page Reference:

33. In a one-point perspective lines that are perpendicular to the picture plane converge at the vanishing point. True or False?

Answer: True

Question Type: T
Random: Y
Difficulty Level: 2
Chapter Number: 9
Page Reference: 188

34. When showing reflections, how is the vanishing point for the reflection located?

Answer: No vanishing points need to be established, the vanishing points for the perspective should be used.

Question Type: S
Random: Y
Difficulty Level: 3
Chapter Number: 9
Page Reference: 195

35. Discuss perspective charts or grids.

Answer: Perspective charts are grids pre-printed to aid setting up a perspective. These grids have established vanishing points.

Question Type: S
Random: Y
Difficulty Level: 2
Chapter Number: 9
Page Reference: 196

36. The picture plane line used in plan view, represents what object?

Answer: The drawing media onto which the perspective is being projected.

Question Type: S
Random: Y

Difficulty Level: 3
Chapter Number: 9
Page Reference: 170

Chapters 10. SHADES AND SHADOWS

1. Why is the action of light on architectural subjects important to the designer?

Answer: The action of light on architectural subjects is important to the designer. Good architectural forms have the property of producing pleasing shadows regardless of the sunups position. To the observer, shadows are an integral part of an architectural composition, and their representation becomes almost as important as the building itself.

Question Type: S
Random: Y
Difficulty Level: 2
Chapter Number: 10
Page Reference: 205

2. Why is it convenient to use a 45° light source direction?

Answer: It is convenient to use a 45. light source direction because it can be easily drawn with a triangle. Another advantage is that it conveniently reveals depth-dimensional characteristics.

Question Type: S
Random: Y
Difficulty Level: 2
Chapter Number: 10
Page Reference: 210

3. Explain the difference between a shade and a shadow.

Answer: Objects in light cast shadows: objects not receiving light are in shade.

Question Type: S
Random: Y
Difficulty Level: 2
Chapter Number: 10
Page Reference: 210

4. What purpose do orthographic shadows serve?

Answer: Orthographic shadows show depth information by association with related orthographic views.

Question Type: S
Random: Y
Difficulty Level: 2
Chapter Number: 10
Page Reference: 206

5. When the light source is parallel to the picture plane, is the shadow of a vertical corner drawn vertically when it falls on a wall? Explain.

Answer: Yes, when the light source is parallel to the picture plane, the shadow of a vertical corner is drawn vertically when it falls on a wall. Explain.

Question Type: S
Random: Y
Difficulty Level: 2
Chapter Number: 10
Page Reference: 209

6. When drawing lines to represent shadows falling on a ground plane, would it be best to draw the lines vertically or horizontally?

Answer: When drawing lines to represent shadows falling on a ground plane, it would be best to draw the lines horizontally.

Question Type: S
Random: Y
Difficulty Level: 2
Chapter Number: 10
Page Reference: 210

7. The conventional light source in orthographic shades and shadows is assumed to be the diagonal of a cube. True or False?

Answer: True

Question Type: T
Random: Y
Difficulty Level: 1
Chapter Number: 10
Page Reference: 205

8. Shading pictorial sketches help define the masses. True or False?

Answer: True

Question Type: T
Random: Y
Difficulty Level: 1
Chapter Number: 10
Page Reference: 214

9. To add more definition to the exterior of a building, the shades and shadows can be developed by adding additional light sources. True or False?

Answer: False

Question Type: T
Random: Y
Difficulty Level: 2
Chapter Number: 10
Page Reference: 205

10. What are the three (3) conditions that will affect a shadow?

1. ________________________

2. ________________________

3. ________________________

Answer: 1. The direction of the light source.
2. The shape of the object.
3. The manner and shape of the surface upon which the shadow falls.

Question Type: F
Random: Y
Difficulty Level: 2
Chapter Number: 10
Page Reference: 207

11. Name six general consistencies of shadows.

1. ____________________

2. ____________________

3. ____________________

4. ____________________

5. ____________________

6. ____________________

Answer: 1. Only an object in light casts a shadow.
2. A shadow is cast only when it falls on a lighted surface.
3. A shadow of a point must be on the light ray of that point.
4. Parallel surfaces cast parallel shadows.
5. Figures cast identical shadows on parallel surfaces.
6. A shadow of a line perpendicular to the picture plane will be inclined on a surface parallel to the picture plane.

Question Type: F
Random: Y
Difficulty Level: 3
Chapter Number: 10
Page Reference: 207

12. In perspective shades and shadows, the shadow cast by a vertical line on a horizontal plane is __________________.

Answer: Horizontal

Question Type: F
Random: Y
Difficulty Level: 1
Chapter Number: 10
Page Reference: 210

13. In perspective shades and shadows, on parallel surfaces, a shadow is parallel to the line that cast it. Therefore, it should be taken to ____________________________.

Answer: The vanishing point.

Question Type: S
Random: Y
Difficulty Level: 3
Chapter Number: 10
Page Reference: 210

14. Parallel surfaces cast perpendicular shadows. True or False?

Answer: False

Question Type: T
Random: Y

Difficulty Level: 1
Chapter Number: 10
Page Reference: 207

15. In perspective shades and shadows, the shadow of a vertical line is ____________________ if it falls on a vertical surface.

Answer: vertical

Question Type: F
Random: Y
Difficulty Level: 1
Chapter Number: 10
Page Reference: 210

16. In perspective shades and shadows, the shadow of a horizontal line is ____________________ if it falls on a vertical surface.

Answer: Inclined.

Question Type: F
Random: Y
Difficulty Level: 3
Chapter Number: 10
Page Reference: 210

Chapter 11. PRESENTATION DRAWINGS AND RENDERINGS

1. What are the advantages for clients and architects, of preparing a presentation drawing of an architectural proposal?

Answer: There are several advantages for a client and an architect to preparing a presentation drawing of an architectural proposal. These include providing the client with both a tangible proposal for approval and also a marketing instrument. It also offers a means for careful study of the structures appearance so that any changes and improvements may be made.

Question Type: S
Random: Y
Difficulty Level: 1
Chapter Number: 11
Page Reference: 223

2. For what reason are perspective views generally effective in presentation work?

Answer: Perspective views are generally effective because they present the building in its natural setting in an honest, realistic, artistic manner which is easily understood by the layperson.

Question Type: S
Random: Y
Difficulty Level: 2
Chapter Number: 11
Page Reference: 223

3. What media are appropriate for rendering small drawings?

Answer: The most frequently used media for rendering small drawings are pencil or pen and ink.

Question Type: S
Random: Y
Difficulty Level: 2
Chapter Number: 11
Page Reference:

4. On perspective renderings, why should the details nearest the observer be made more prominent than those in the distance?

Answer: In reality near objects are sharper and closer to the eye than those in the distance.

Question Type: S
Random: Y
Difficulty Level: 1
Chapter Number: 11
Page Reference: 231

5. What do landscaping indications contribute to a perspective?

Answer: Landscaping indications in a perspective tie elements together. They tend to keep strong, isolated elements from falling away from the layout. They also give the drawing realism.

Question Type: S
Random: Y
Difficulty Level: 1
Chapter Number: 11
Page Reference: 238

6. What is meant by a value study?

Answer: The value study is achieved by tracing over the completed perspective and establishing the general value patterns rather roughly, using a soft pencil or charcoal. Avoid details in this study, but strive for pleasing value patterns that can be created by contrasts of light and shade and from the landscaping composition. This becomes your guide for choosing line technique, value, and landscaping on the final ink drawing.

Question Type: E
Random: Y
Difficulty Level: 2
Chapter Number: 11
Page Reference: 240

7. How are analogous colors related to each other on the color wheel?

Answer: Analogous colors are adjacent to each other on the color wheel.

Question Type: S
Random: Y
Difficulty Level: 2
Chapter Number: 11
Page Reference: 246

8. When rendering sky, should the sky appear darker near the building or near the top edge of the rendering.

Answer: When rendering sky, the sky should appear darker near the edge of the rendering.

Question Type: S
Random: Y
Difficulty Level: 1
Chapter Number: 11
Page Reference: 231

9. What is the first step you should take in drawing a tree?

Answer: The first step that should be taken in drawing a tree is direct and careful observation from nature. Draw the structure of the tree by drawing the trunk and the branches before adding foliage.

Question Type: S
Random: Y
Difficulty Level: 2
Chapter Number: 11
Page Reference: 238

10. List three types of opaque water colors frequently used in renderings.

Answer: Three types of opaque water colors frequently used in rendering are: tempera, acrylic, or casein.

Question Type: S
Random: Y
Difficulty Level: 2
Chapter Number: 11
Page Reference: 247

11. People drawn on a rendering help the viewer to gauge size by comparing the height of the person to the building. True or False?

Answer: True

Question Type: T
Random: Y
Difficulty Level: 1
Chapter Number: 11
Page Reference: 247

12. What drawing medium would be appropriate if a rendering must be prepared very quickly?

Answer: When a rendering must be done very quickly, the most appropriate medium would be felt tip pens.

Question Type: S
Random: Y
Difficulty Level: 2
Chapter Number: 11
Page Reference: 227

13. Discuss ways to achieve contrast in a rendering.

Answer: The variation of lightness or darkness of color or gray is known as value. Before starting a drawing, develop a value scale that ranges from black to white. Plan the drawing so that important features are accented with extreme contrasts and less important features or outlines have lesser amounts of contrast. Plan a definite arrangement of the values.

Question Type: E
Random: Y
Difficulty Level: 2
Chapter Number: 11
Page Reference: 231

14. Why is waterproof ink essential when it is combined with a wet medium?

Answer: Waterproof ink is essential when combined with a wet medium to prevent smearing, running and smudging.

Question Type: S
Random: Y
Difficulty Level: 1
Chapter Number: 11
Page Reference: 240

15. What would be the advantage of preparing a pencil or ink rendering on vellum instead of illustration board?

Answer: The advantage of preparing a pencil or ink rendering on vellum instead of illustration board would be the ability to reproduce copies from the original vellum or Mylar by the diazo or photo copy process.

Question Type: S
Random: Y
Difficulty Level: 1
Chapter Number: 11
Page Reference: 227

16. What are two (2) reasons why presentation drawings are prepared?

Answer: Presentation drawings are prepared because (1) it combines the efforts of the preliminary planning into a tangible proposal for a client's approval and (2) it offers a means for careful study of the structure's appearance upon which improvements or changes for final development can be based.

Question Type: S
Random: Y
Difficulty Level: 1
Chapter Number: 11
Page Reference: 223

17. What are four of the different media that may be used in effective presentation drawings?

Answer: The different media that may be used in effective presentation drawings are (1) pencil, (2) pen and ink, (3) colored pencil, (4) water color, (5) tempera, (6) acrylic, (7) pastel and (8) charcoal pencil. These media may be used individually or in combination with each other.

Question Type: S
Random: Y
Difficulty Level: 2
Chapter Number: 11
Page Reference:

18. Why are plumbing fixtures and furniture usually shown on floor plan presentation drawings?

Answer: Plumbing fixtures and furniture are usually shown on floor plan presentation drawings to indicate the workability of the design, the space around these items and allow the client to determine whether his or her space requirements have been met.

Question Type: S

Random: Y
Difficulty Level: 2
Chapter Number: 11
Page Reference:

19. Pencil drawings may be done on ______________________________ or

on ____ .

Answer: A good grade tracing paper or on illustration board.

Question Type: F
Random: Y
Difficulty Level: 1
Chapter Number: 11
Page Reference: 227

20. What are some of the advantages of using tracing paper for presentation work?

Answer: Some of the advantages of using tracing paper for presentation work are (1) it is convenient for reproduction, (2) it aids trial and error composition with little wasted effort and (3) weak points of the composition, when viewed from the reverse side, can be quickly observed.

Question Type: S
Random: Y
Difficulty Level: 2
Chapter Number: 11
Page Reference: 227

21. What methods can the renderer use to produce emphasis?

Answer: The renderer can emphasize areas of the presentation by (1) adjusting the angle of the perspective, (2) using the contrast of dark and light to make areas more dominant and (3) use foliage or other objects to carry the observers eye to the feature to be emphasized.

Question Type: S
Random: Y
Difficulty Level: 2
Chapter Number: 11
Page Reference: 231

22. Define "composition" as it relates to renderings.

Answer: Composition is the art of arranging lines, values, spaces, masses and other elements into a thing of beauty.

Question Type: S
Random: Y
Difficulty Level: 2
Chapter Number: 11
Page Reference: 227

23. What does landscaping and indications of people add to a presentation?

Answer: Landscaping softens the hard mechanical lines of man-made objects by bringing natural items into the composition and can, if properly located, emphasize areas. People add scale to a presentation by giving the observer a size and dimension with which he can associate.

Question Type: E
Random: Y
Difficulty Level: 2
Chapter Number: 11
Page Reference: 240

24. Name three (3) things to avoid when rendering a presentation.

1. ______________________________
2. ______________________________
3. ______________________________

Answer:
1. Mechanical and uniform patterns unless repetition is necessary for harmony.
2. Strong lines near and parallel to the edge of the paper or board.
3. Two or three strong elements in a straight line.

Question Type: F
Random: Y
Difficulty Level: 2
Chapter Number: 11
Page Reference: 228

25. What is meant by "optical balance"?

Answer: Each element of a rendering has optical weight, i.e., large over small, heavier over lighter, bright over pale, etc. and these graphic weights must balance or the composition will become one sided or top/bottom heavy. The balancing of these weights is called optical balance.

Question Type: E
Random: Y
Difficulty Level: 2
Chapter Number: 11
Page Reference: 230

26. Presentations, in order to have the most effect, must be highly detailed and indicate all items of the object as they will appear in the final product. True or False?

Answer: False

Question Type: T
Random: Y
Difficulty Level: 2
Chapter Number: 11
Page Reference: 231

27. How can a sky be given a dramatic effect in a rendering?

Answer: A dramatic sky can be created in a rendering by making the sky darker at the building and fading it out towards the edge of the paper or board. With the building light in contrast and some minor clouds, a very dramatic sky can be attained.

Question Type: E
Random: Y
Difficulty Level: 3
Chapter Number: 11
Page Reference: 231

28. What pencils are recommended for pencil renderings?
a. ______________________
b. ______________________
c. ______________________
d. ______________________
e. ______________________

Answer: a: " H" b: "HB" c: "2B" d: "4B" e: "6B"

Question Type: F
Random: Y
Difficulty Level: 2

Chapter Number: 11
Page Reference: 232

29. What are the two (2) types of pencil points used in rendering?

Answer: 1. Cone shaped for outlines and fine pencil work. 2. Chisel points for laying out various tones - sometimes called "pencil painting".

Question Type: F
Random: Y
Difficulty Level: 2
Chapter Number: 11
Page Reference: 233

30. Why is the pencil more suitable than pen and ink for rendering different textures?

Answer: The pencil is a medium that lends itself well to soft, tonal variations and therefore is more suitable than pen and ink for the rendering of textures.

Question Type: S
Random: Y
Difficulty Level: 2
Chapter Number: 11
Page Reference: 232

31. Why can stonework be rendered either rough or smooth?

Answer: Stonework can be rendered either rough or smooth because cut stone usually has uniform coursing and a rather smooth surface whereas rubble stonework may have various unit sizes and a rough and irregular texture.

Question Type: S
Random: Y
Difficulty Level: 2
Chapter Number: 11
Page Reference: 233

32. Plaster, stucco and E.I.F.S. construction may be represented on the rendering by using a ______________________ technique.

Answer: Dots or stippled.

Question Type: F
Random: Y
Difficulty Level: 2
Chapter Number: 11
Page Reference: 243

33. Why are trees an important addition to a rendering?

Answer: Trees are an important addition to a rendering because, when drawn in the correct proportion to a building, establish the relative scale in the viewer's mind and add a note of softness and charm to the rendering.

Question Type: S
Random: Y
Difficulty Level: 2
Chapter Number: 11
Page Reference: 238

34. What individual characteristics of a tree should students look for when familiarizing themselves with a particular species?

Answer: Students should be aware of the tree's silhouette, trunk and branch structure, the bark texture, the appearance at the base of the trunk and the foliage clusters and leaf forms.

Question Type: S
Random: Y
Difficulty Level: 2
Chapter Number: 11
Page Reference: 238

34. What is the purpose of indicating shrubbery near the building?

Answer: The purpose of indicating shrubbery near the building help to tie the building and ground together and eliminate hard vertical architectural lines.

Question Type: S
Random: Y
Difficulty Level: 2
Chapter Number: 11
Page Reference: 239

36. Trees and shrubbery on a rendering should be subordinate to the architecture. True or False?

Answer: True

Question Type: T
Random: Y
Difficulty Level: 2
Chapter Number: 11
Page Reference: 238-9

37. The variation of lightness or darkness of color or grays is known as ______________________________ .

Answer: Value.

Question Type: F
Random: Y
Difficulty Level: 1
Chapter Number: 11
Page Reference: 240

38. Define "contrast".

Answer: Contrast is defined as the striking difference in adjacent elements of a composition, whether by the use of color, value, texture, outline or treatment.

Question Type: S
Random: Y
Difficulty Level: 1
Chapter Number: 11
Page Reference: 231

39. What are the basic materials required when rendering in pen and ink?

Answer: The basic materials required when rendering in pen and ink are (1) a strong paper or board that will take ink well, (2) black waterproof India ink, (3) various size technical drafting pens, (4) crow quill pen points and holder, (5) a Dubbed point pen and holder, (6) a small pointed watercolor brush and (7) a small bottle of white tempera color.

Question Type: S
Random: Y
Difficulty Level: 2
Chapter Number: 11
Page Reference: 240

40. Why is a "value study" recommended prior to starting a pen and ink rendering?

Answer: A value study is recommended prior to starting a pen and ink rendering since tonal variations are difficult with this medium and the placement of the various tonal qualities must be determined before the work may commence.

Question Type: S
Random: Y
Difficulty Level: 3
Chapter Number: 11
Page Reference: 242

41. Primary colors are ______________________.

Answer: Primary colors are red, blue and yellow.

Question Type: F
Random: Y
Difficulty Level: 1
Chapter Number: 11
Page Reference: 246

42. Secondary colors are ____________________ .

Answer: Secondary colors are orange, green and purple.

Question Type: F
Random: Y
Difficulty Level: 1
Chapter Number: 11
Page Reference: 246

43. Tertiary colors are __________________ .

Answer: Tertiary colors are those colors located between the primary and secondary colors on the color wheel and are made by mixing adjacent colors.

Question Type: F
Random: Y
Difficulty Level: 1
Chapter Number: 11
Page Reference: 246

44. Warm colors seem to advance towards the viewer while the cool colors appear to recede when seen from the same distance. True or False?

Answer: True

Question Type: T
Random: Y
Difficulty Level: 1
Chapter Number: 11
Page Reference: 246

45. What is a "bridge" and how is it used when preparing tempera renderings.

Answer: A bridge is a rigid tube or slat with blocks fastened at each end to raise the tube above the surface of the rendering. Its purpose is to allow the renderer to paint straight lines on a building without touching the drawing. The clearance of the tube above the paper or board also allows the wet paint to be applied without smearing.

Question Type: S
Random: Y
Difficulty Level: 2
Chapter Number: 11
Page Reference: 249

46. What are the four (4) basic paints used on renderings?

1. ______________________________

2. ______________________________

3. ______________________________

4. ______________________________

Answer: 1. Water colors 2. Casein 3. Tempera 4. Acrylic

Question Type: F
Random: Y
Difficulty Level: 2
Chapter Number: 11
Page Reference: 247

47. Drawing mediums that would be appropriate if a rendering is to be

done quickly are ____________, and ____________.

Answer: Pencils and colored pencil.

Question Type: F
Random: Y
Difficulty Level: 2
Chapter Number: 11
Page Reference:

48. Why are colored pencils a good media for the beginner?

Answer: Colored pencils are a good media for the beginner since there is an extensive range of colors available and they may be erased if changes must be made.

Question Type: S
Random: Y
Difficulty Level: 2
Chapter Number: 11
Page Reference: 246

49. When producing quick sketch renderings on tracing paper, the architect usually uses ______________________.

Answer: Felt-tip pens

Question Type: F
Random: Y
Difficulty Level: 2
Chapter Number: 11
Page Reference: 249

50. The two (2) tips available for felt-tip renderings are _________ and ______________.

Answer: Cone and chisel.

Question Type: F
Random: Y
Difficulty Level: 2
Chapter Number: 11
Page Reference: 249

51. ______________ is used to designate the name of a color.

Answer: Hue

Question Type: F
Random: Y
Difficulty Level: 2
Chapter Number: 11
Page Reference: 246

52. ________________ is intensity or purity of a color.

Answer: Chorma

Question Type: F
Random: Y
Difficulty Level: 2
Chapter Number: 11
Page Reference: 246

53. ________________ is the darkened value of a color.

Answer: Shade

Question Type: F
Random: Y
Difficulty Level: 2
Chapter Number: 11
Page Reference: 246

54. ________________ is the lightened value of a color.

Answer: Tint

Question Type: F
Random: Y
Difficulty Level: 2
Chapter Number: 11
Page Reference: 246

55. ________________ is a color scheme whereby values of only one color is used.

Answer: Monochromatic

Question Type: F
Random: Y
Difficulty Level: 2
Chapter Number: 11
Page Reference: 246

56. Complementary colors are located __________________ on the color wheel.

Answer: Opposite each other.

Question Type: F
Random: Y
Difficulty Level: 2
Chapter Number: 11
Page Reference: 246

57. Primary colors are those colors use most often. True or False?

Answer: False

Question Type: T
Random: Y
Difficulty Level: 1
Chapter Number: 11
Page Reference: 246

58. Warm colors are:
1. ____________
2. ____________
3. ____________

Answer: 1. Reds 2. Oranges 3. Yellows

Question Type: F
Random: Y
Difficulty Level: 2
Chapter Number: 11
Page Reference: 246

59. Cold colors are:
1. ____________
2. ____________
3. ____________

Answer: 1. Blues 2. Greens 3. Violets

Question Type: F
Random: Y
Difficulty Level: 2
Chapter Number: 11
Page Reference: 246

MULTIPLE CHOICE: Circle the letter indicating the best choice for each question.

60. When determining value arrangements which produce the most interesting shadow patterns, generally:
    a. Features in the foreground can be made the darkest.
    b. Features in mid-distance can be of intermediate tones.
    c. Features in the distance can be the lightest.
    d. All of the above.

Answer: d

Question Type: M
Random: Y
Difficulty Level: 2
Chapter Number: 11
Page Reference: 242

61. The most effective method for rendering glass is:
    a. Leave it transparent.
    b. Render dark but leave highlights of reflection on the glass.
    c. Render totally dark from top to bottom of sash.
    d. Show reflections of other objects in the glass.
    e. All of the above.

Answer: b

Question Type: M
Random: Y
Difficulty Level: 2
Chapter Number: 11
Page Reference: 236

Chapter 12. PRINCIPLES OF LIGHT CONSTRUCTION

1. Why must local frost depths be considered when drawing footings?

Answer: Local frost depths are prescribed to prevent damage to the footings from frost action such as freezing and thawing that could cause failure of concrete.

Question Type: S
Random: Y
Difficulty Level: 1
Chapter Number: 12
Page Reference: 257

2. The Uniform Building Codes prescribe footings to be place a minimum of ______________ below the frost line.

Answer: 1'-0".

Question Type: F
Random: Y
Difficulty Level: 1
Chapter Number: 12
Page Reference: 258

3. Describe a stepped footing and indicate its application.

Answer: Stepped footings are usually needed on sloping or steep lot. where straight horizontal footings would be uneconomical. stepped footings change planes to follow the contour of the site.

Question Type: E
Random: Y
Difficulty Level: 2
Chapter Number: 12
Page Reference: 275

4. What conditions may require accurate calculation of footing size?

Answer: The soil quality and moisture content will require accurate calculations of footing sizes.

Question Type: S
Random: Y

Difficulty Level: 2
Chapter Number: 12
Page Reference: 258

5. What is a pier?

Answer: A pier is a freestanding post usually constructed of masonry, concrete or pressure treated wood and is used to support a floor beam.

Question Type: S
Random: Y
Difficulty Level: 2
Chapter Number: 12
Page Reference: 276

6. What are the advantages of a slab-on-ground floor construction in comparison to crawl space construction?

Answer: The advantages of slab on-ground construction over crawl space construction are outdoor surfaces become more accessible from the interior of the house, a lower silhouette in the overall appearance of the house is possible and the concrete slab may be more cost effective than wood floor framing in some areas.

Question Type: S
Random: Y
Difficulty Level: 2
Chapter Number: 12
Page Reference: 270-272

7. Indicate the logical application of grade beam foundation construction.

Answer: Grade beam foundation construction is used to support light frame structures in areas where poor soil is encountered.

Question Type: S
Random: Y
Difficulty Level: 2
Chapter Number: 12
Page Reference: 277

8. What are the identifying characteristics of platform (western) frame construction?

Answer: The characteristics that identify platform frame construction are one floor level is built at a time, thus permitting the completed floor surface to serve as a platform on which walls and the like can be constructed.

Question Type: S
Random: Y
Difficulty Level: 2
Chapter Number: 12
Page Reference: 284

9. Why are roof trusses widely used in small home construction?

Answer: Trusses are widely used because they are strong and are designated to clear span a residence eliminating the need for interior bearing walls. They are usually prefabricated and quickly installed.

Question Type: S
Random: Y
Difficulty Level: 2
Chapter Number: 12
Page Reference: 324

10. In plank and beam framing, how is the load on a roof beam transmitted to the foundation?

Answer: The load on roof beams is transferred directly to the foundation by the use of posts or columns.

Question Type: S
Random: Y
Difficulty Level: 1
Chapter Number: 12
Page Reference: 326

11. Why should the location of beams in plank and beam framing be shown on the working drawings?

Answer: The location of beams on working drawings and beam framing is critical no that the beam might be precisely placed during the construction thus allowing the beam to transmit loads properly.

Question Type: S
Random: Y
Difficulty Level: 2
Chapter Number: 12

Page Reference: 326

12. What part of the construction limits the spacing of beams in plank and beam framing?

Answer: Floor and roof loans limit the spacing of beams in plank and beam framing. The allowable span of wood planks may also limit the spacing of beams that support the planks.

Question Type: S
Random: Y
Difficulty Level: 2
Chapter Number: 12
Page Reference: 326

13. What is a transverse beam?

Answer: A transverse beam is a beam that span. from the ridge beam to the exterior wall and rests on a post or column. A transverse beam is usually shorter in length than a longitudinal beam.

Question Type: S
Random: Y
Difficulty Level: 1
Chapter Number: 12
Page Reference: 330

14. List the advantages of brick veneer construction.

Answer: The advantages of brick veneer construction are: durability, low maintenance, long-range economy, and availability in a wide variety of colors.

Question Type: S
Random: Y
Difficulty Level: 2
Chapter Number: 12
Page Reference: 334

15. Explain modular sizing of concrete block walls.

Answer: Modular sizing is based on measurements that are multiples of 4". The joint centerline between concrete blocks is an even 16" wide and 8" high: thus, it is easy to construct concrete block walls in modular sizes.

Question Type: S
Random: Y
Difficulty Level: 2
Chapter Number: 12
Page Reference: 279

16. What is the purpose of a vapor barrier in a wood wall?

Answer: Vapor barriers are used in wood walls to keep condensation out of the wall. They are placed next to the warm humid side of the wall for the purpose of preventing vapor-laden warm air from reaching the insulation in the wall and condensing.

Question Type: E
Random: Y
Difficulty Level: 2
Chapter Number: 12
Page Reference: 385

17. What is flashing and identify two places where flashing is used?

Answer: Flashing is usually thin metal that is used to shed moisture away from a joist or the intersection of dissimilar materials. Flashing is placed between a chimney and a roof, above a window or door, and through a cavity wall to divert water to the exterior.

Question Type: S
Random: Y
Difficulty Level: 2
Chapter Number: 12
Page Reference: 363

18. ________________, ________________ and ________________ are terms given to horizontal structural members used in construction.

Answer: Girders, beams and joists

Question Type: F
Random: Y
Difficulty Level: 1
Chapter Number: 12
Page Reference: 257

19. What is the function of the "footing"?

Answer: The function of the footing is to support the weight of the building, sustain the structure throughout all weather conditions, distribute the weight of the building to the ground below and be deep enough to prevent frost action.

Question Type: E
Random: Y
Difficulty Level: 1
Chapter Number: 12
Page Reference: 257

20. Ideally, footings should be placed on ____________________.

Answer: Undisturbed soil

Question Type: F
Random: Y
Difficulty Level: 1
Chapter Number: 12
Page Reference: 258

21. Residential foundation walls are usually made of _______________, _______________________________ or _______________________________.

Answer: 1. reinforced concrete 2. horizontally reinforced concrete masonry units (CMU) 3. Treated wood.

Question Type: F
Random: Y
Difficulty Level: 2
Chapter Number: 12
Page Reference: 264

22. Describe the dampproofing of a residential CMU foundation wall.

Answer: A residential CMU foundation wall is dampproofed by applying a layer (usually 1/2" thick) of cement pargeting to the exterior of the wall and providing a layer of bituminous dampproofing over the parging.

Question Type: E
Random: Y
Difficulty Level: 3
Chapter Number: 12
Page Reference: 277

23. Why are foundation drains used around residential foundation walls?

Answer: Foundation drains are used around residential foundation walls to collect and dispose of ground water away from the building, thus reducing hydrostatic pressure and the possibility of foundation leaks.

Question Type: E
Random: Y
Difficulty Level: 2
Chapter Number: 12
Page Reference: 265

24. In masonry construction what is a lintel, its purpose and material?

Answer: A lintel is a structural member which spans an opening and supports the masonry over the opening. It may be pre-cast reinforced concrete, individual steel members or pre-cast reinforced concrete masonry units

Question Type: S
Random: Y
Difficulty Level: 2
Chapter Number: 12
Page Reference: 335

25. What is an areaway?

Answer: An areaway is a concrete, masonry or metal retaining device used at below or near grade basement windows to allow ventilation and natural lighting into the basement.

Question Type: S
Random: Y
Difficulty Level: 1
Chapter Number: 12
Page Reference: 280

26. Where would a concrete slab-on-ground be used?

Answer: A concrete slab-on-ground would be used in a basement-less building located on a nearly level lot requiring a minimum of excavation and one free of moisture problems.

Question Type: S
Random: Y
Difficulty Level: 2
Chapter Number: 12
Page Reference: 272

27. What is perimeter insulation and what is its purpose?

Answer: Perimeter insulation is a permanent, waterproof, non-absorptive type of rigid insulation installed around the perimeter of the building below the concrete slab-on-ground. Its purpose is to reduce the heat loss from the slab-on-ground through the ground to the exterior.

Question Type: S
Random: Y
Difficulty Level: 2
Chapter Number: 12
Page Reference: 275

28. How are steel reinforcing rods sized?

Answer: Steel reinforcing rods are sized according to the diameter of the rod in 1/8 inch increments.

Question Type: S
Random: Y
Difficulty Level: 2
Chapter Number: 12
Page Reference: 281

29. What are the three (3) methods used to preserve the maximum service life of wood?

1. ______________________________

2. ______________________________

3. ______________________________

Answer: 1. Control of the moisture content 2. Providing effective insect and termite barriers 3. Using naturally durable or chemically treated wood in critical places in the structure.

Question Type: F
Random: Y
Difficulty Level: 2

Chapter Number: 12
Page Reference: 283

30. What are five (5) advantages of platform framing in light frame construction?

1. ______________________________

2. ______________________________

3. ______________________________

4. ______________________________

5. ______________________________

Answer: 1. Short members are more readily available than longer members required for balloon framing 2. The platform is convenient for carpenters to work upon. 3. The construction has natural "fire-stops" built in 4. It lends itself to modern pre-fabrication methods 5. It is quicker to erect

Question Type: F
Random: Y
Difficulty Level: 2
Chapter Number: 12
Page Reference: 284

31. Discuss wood floor trusses and their advantages in light frame construction.

Answer: Wood floor trusses are an alternative to solid wood joists and are factory assembled units fabricated from 2 x 4 structural grade lumber composed of a top chord, a bottom chord and web members connected by stamped metal plates. Wood floor trusses are capable of spanning greater distances than solid wood joists and are used where large open spaces are required on the floor below. Wiring and ductwork may be routed through the openings in the web.

Question Type: E
Random: Y
Difficulty Level: 3
Chapter Number: 12
Page Reference: 298

32. Define a "trimmer", as it applies to a floor system.

Answer: A trimmer is a doubled joist running parallel to the two sides of an opening in a floor.

Question Type: S
Random: Y
Difficulty Level: 1
Chapter Number: 12
Page Reference: 294

33. Define a "header", as it applies to a floor system.

Answer: A header is the doubled member at the end of an opening in a floor. The header is connected to the trimmers at either end and carries the shortened joists called header joists or tail joists.

Question Type:
Random: Y
Difficulty Level: 1
Chapter Number: 12
Page Reference: 294

34. Discuss "bridging" and its purpose.

Answer: Bridging consists of rows of short members of wood or metal placed in the form of an "X" or solid wood members placed between floor or ceiling joists to produce a firm, rigid platform. Bridging prevents the deep narrow shape of the floor joist from buckling sideways and distributes concentrated live loads to other joists.

Question Type: S
Random: Y
Difficulty Level: 1
Chapter Number: 12
Page Reference: 297

35. What is the purpose of the sub-floor?

Answer: The purpose of the sub-floor is to add rigidity to the platform, provide a working floor for the carpenters and provide a level, flat, nailable surface for the attachment of the finished floor.

Question Type: S
Random: Y

Difficulty Level: 2
Chapter Number: 12
Page Reference: 304

36. What is the purpose of sheathing in light frame construction?

Answer: The purpose of sheathing in light frame construction is to add rigidity to the vertical walls - especially at the corners, to help insulate the building and to provide a surface to attach the exterior cladding.

Question Type: S
Random: Y
Difficulty Level: 3
Chapter Number: 12
Page Reference: 319

37. Roof rafters must be securely anchored on the leeward side as well as the windward side of the building. True or False?

Answer: True

Question Type: T
Random: Y
Difficulty Level: 2
Chapter Number: 12
Page Reference: 312

38. What is a "valley rafter"?

Answer: A valley rafter is the rafter directly below the intersection of two sloping roofs.

Question Type: S
Random: Y
Difficulty Level: 1
Chapter Number: 12
Page Reference: 313

39. What is a "jack rafter"?

Answer: A jack rafter is the shortened rafter that runs from the hip rafter to the top plate or from the valley rafter to the ridge.

Question Type: S
Random: Y

Difficulty Level: 1
Chapter Number: 12
Page Reference: 313

40. Describe a pre-fabricated roof truss and discuss the reasons for its use in light frame construction.

Answer: A pre-fabricated roof truss is a factory made system of triangular shapes designed to span a given distance with a given roof slope. The advantages of this system over solid wood framing is that the roof truss is complete, uniform and easily installed in less time than the individual "stick" framing. The roof truss may be made from smaller members since all members work together in carrying the loads whereas the "stick" framing members work individually.

Question Type: E
Random: Y
Difficulty Level: 3
Chapter Number: 12
Page Reference: 324

41. Discuss the "sill plate" in balloon framing.

Answer: The purpose of the sill plate in balloon framing is to provide a member to fasten the upper frame to the foundation wall. The sill plate must be pressure treated against insects, moisture and decay and be fastened to the foundation wall by a series of anchor bolts.

Question Type: E
Random: Y
Difficulty Level: 2
Chapter Number: 12
Page Reference: 322

42. What is a "longitudinal" beam?

Answer: In plank and beam construction a longitudinal beam is a beam that runs the entire length of the building and supports the roof decking spanning the distance from one longitudinal beam to another.

Question Type: S
Random: Y
Difficulty Level: 1
Chapter Number: 12

Page Reference: 331

43. Discuss the advantages of glue-laminated members.

Answer: Glue-laminated members have many advantages over solid wood members. Because of the conditions of their manufacture the moisture content of the wood is controlled giving stability in size, shape and movement due to shrinkage and swelling. Since the glue-lams are composed of individual small members, various and unorthodox shapes may be manufactured, large wood members are not required and members of any size may be obtained.

Question Type: E
Random: Y
Difficulty Level: 2
Chapter Number: 12
Page Reference: 302

44. The soldier course exposes the ________________________.

a. The face of the brick in a vertical position.
b. The face of the brick in a horizontal position.
c. The head of the brick in a vertical position.
d. The head of the brick in a horizontal position.
e. The bed of the brick in a vertical position.
f. The bed of the brick in a horizontal position.

Answer: a

Question Type: M
Random: Y
Difficulty Level: 2
Chapter Number: 12
Page Reference: 336

45. What is a "fire-cut" and why is it necessary?

Answer: A fire-cut is the diagonal cutting of the end of a wood member bearing in a masonry wall. Its purpose is to allow the wood member to rotate out of the pocket in the masonry wall without damaging the masonry wall if the wood member fails at its center.

Question Type: S
Random: Y
Difficulty Level: 2
Chapter Number: 12

Page Reference: 337

46. What is the purpose of horizontal steel reinforcing in masonry walls?

Answer: The purpose of horizontal steel reinforcing in masonry walls is to attempt to eliminate cracking in the wall.

Question Type: S
Random: Y
Difficulty Level: 2
Chapter Number: 12
Page Reference: 339

47. What are the important factors in the selection of roofing material?

Answer: The important factors in the selection of roofing material are 1) the slope of the roof, 2) the quality or permanence required, 3) the inherent architectural features of the roofing in relation to the rest of the structure and 4) the cost of the roofing material.

Question Type: S
Random: Y
Difficulty Level: 3
Chapter Number: 12
Page Reference: 360

48. What are the three (3) basic types of roofing systems?

1. ______________________________

2. ______________________________

3. ______________________________

Answer: 1. Built-up roofing 2. Membrane roofing 3. Shingles

Question Type: F
Random: Y
Difficulty Level: 2
Chapter Number: 12
Page Reference: 360-2

49. What roofing materials could be used on a sloping roof?

1. ______________________________

2. ______________________________

3. ______________________________

4. ______________________________

5. ______________________________

6. ______________________________

7. ______________________________

8. ______________________________

Answer: 1. Asphalt shingles 2. Fiberglass shingles 3. Wood shingles 4. Wood shakes 5. Slate shingles 6. Clay tile units 7. Cement tile units 8. Metal shingles or panels

Question Type: F
Random: Y
Difficulty Level: 3
Chapter Number: 12
Page Reference: 363

50. What are the three (3) basic types of seams used in metal roofing systems?

1. ______________________________

2. ______________________________

3. ______________________________

Answer: 1. Batten 2. Flat 3. Standing

Question Type: F
Random: Y
Difficulty Level: 1
Chapter Number: 12
Page Reference: 372

51. Describe an E.I.F.S. exterior wall treatment system.

Answer: The exterior insulation and finish system (E.I.F.S.) is a four step process of applying a stucco appearing finish to a building. The system includes an exterior insulating panel applied to a masonry wall or gypsum sub-strate protected by an adhesive base coat and an acrylic polymer finish coat. The finish coat may be tinted. The E.I.F.S. treatment adds insulating value to the wall system, will not crack like plaster stucco and is a low maintenance material.

Question Type: E
Random: Y
Difficulty Level: 3
Chapter Number: 12
Page Reference: 377

52. On which side of an exterior wall should the vapor diffusion retarder be placed?

Answer: The vapor retarder should be placed on the warm side (interior) of the exterior wall.

Question Type: S
Random: Y
Difficulty Level: 2
Chapter Number: 12
Page Reference: 385

53. The diameter of a #5 rebar is ________________.

Answer: 5/8" diameter.

Question Type: F
Random: Y
Difficulty Level: 1
Chapter Number: 12
Page Reference: 281

54. Three (3) methods of heat transmittal are:
1. ______________________
2. ______________________
3. ______________________

Answer: 1. Conduction 2. Convection 3. Radiation.

Question Type: F
Random: Y
Difficulty Level: 2
Chapter Number: 12
Page Reference: 380

55. Two State codes, and their reasons, that they require additional reinforcing in foundations are ______________________________ and ________________________ .

Answer: Florida's Hurricane Code and California's Seismic Code.

Question Type: F
Random: Y
Difficulty Level: 1
Chapter Number: 12
Page Reference: 258

56. _______________ loads refer to weights of the structure as well as any stationary equipment.

Answer: Dead.

Question Type: F
Random: Y
Difficulty Level: 1
Chapter Number: 12
Page Reference: 259

57. _______________ loads refer to varying weights and from forces to which parts of a building are subjected.

Answer: Live.

Question Type: F
Random: Y
Difficulty Level: 1
Chapter Number: 12
Page Reference: 259

58. A colorless, odorless, tasteless gas found in soils is _________.

Answer: Radon

Question Type: F
Random: Y
Difficulty Level: 2
Chapter Number: 12
Page Reference: 261

59. Three (3) ways to deal with radon in constructions are:

1. ____________________
2. ____________________

3. ____________________

Answer: 1. Barrier approach
2. Soil gas interception
3. Indoor air management

Question Type: F
Random: Y
Difficulty Level: 2
Chapter Number: 12
Page Reference: 261

60. Three (3) basic types of foundations are:
1. ____________________
2. ____________________
3. ____________________

Answer: 1. Basements
2. Crawl spaces
3. Slab-on-grade

Question Type: F
Random: Y
Difficulty Level: 2
Chapter Number: 12
Page Reference: 264

61. Three (3) methods of preventing water problems in a basement are:
1. ____________________
2. ____________________
3. ____________________

Answer: 1. Surface drainage
2. Subsurface drainage
3. Dampproofing

Question Type: F
Random: Y
Difficulty Level: 2
Chapter Number: 12
Page Reference: 265

62. When the footing and slab are poured at one time, and the footing is a part of the slab, the footing is referred to as a(n) __________________________ footing.

Answer: integral

Question Type: F
Random: Y
Difficulty Level: 2
Chapter Number: 12
Page Reference: 274

63. Foundations made with a system of concrete beams spanning poured concrete piers is a(n) ________________________ foundation.

Answer: grade beam

Question Type: F
Random: Y
Difficulty Level: 2
Chapter Number: 12
Page Reference: 276

64. All-Weather wood foundations are constructed of the same type of wood as the exterior walls. True or False?

Answer: False

Question Type: T
Random: Y
Difficulty Level: 1
Chapter Number: 12
Page Reference: 281

65. Give four (4) advantages of built-up beams.
1. ____________________
2. ____________________
3. ____________________
4. ____________________

Answer: 1. Availability of stock-size lumber
2. Easier to handle smaller pieces
3. Splitting and checking are reduced
4. Nailed to frame more easily

Question Type: F
Random: Y
Difficulty Level: 2
Chapter Number: 12
Page Reference: 289

66. Anchor bolts must extend 24" deep into the concrete foundation and be spaced no further than 2'-0" oc. True or False?

Answer: False

Question Type: T
Random: Y
Difficulty Level: 1
Chapter Number: 12
Page Reference: 289

67. Floor joist should bear a minimum ___________ on the sill plate.

Answer: 1 1/2", preferably 3".

Question Type: F
Random: Y
Difficulty Level: 2
Chapter Number: 12
Page Reference: 291

68. Built-up beams and girders should bear a minimum of ___________ in the foundation beam pocket.

Answer: 3"

Question Type: F
Random: Y
Difficulty Level: 2
Chapter Number: 12
Page Reference: 294

69. Three (3) types of bridging used in a floor system are:
1. ____________________
2. ____________________
3. ____________________

Answer: 1. 1 x 3 diagonal wood
2. Blocking the same depth of floor joist
3. Light metal diagonal

Question Type: F
Random: Y
Difficulty Level: 2
Chapter Number: 12
Page Reference: 20

70. Four(4) groups glued engineered members are:

1. ______________________
2. ______________________
3. ______________________
4. ______________________

Answer: 1. Structural wood panels
2. Glued laminated timber
3. Structural composite lumber (SCL)
4. Wood I-joist

Question Type: F
Random: Y
Difficulty Level: 2
Chapter Number: 12
Page Reference: 302

71. Describe a wood I-joist.

Answer: A wood I-joist is in the shape of an "I" where the flanges are made of either sawn lumber or LVL. The web is made of plywood or OSB.

Question Type: E
Random: Y
Difficulty Level: 2
Chapter Number: 12
Page Reference: 303

72. ______________ is required to restrict movement of flame and gases to other areas of a building through combustible passages.

Answer: Fire blocking

Question Type: F
Random: Y
Difficulty Level: 2
Chapter Number: 12
Page Reference: 308

73. Fire stops shall be provided at these four(4) locations:
1. ______________________
2. ______________________
3. ______________________
4. ______________________

Answer: 1. At 10' intervals both vertically and horizontally in concealed spaces such as stud walls, and partitions (including furred spaces at ceilings and floors)

2. At connections between concealed vertical and horizontal spaces, such as soffits and drop ceilings.
3. In spaces between stair stringers at top and bottom of run.
4. At openings around pipes, ducts, etc.

Question Type: F
Random: Y
Difficulty Level: 3
Chapter Number: 12
Page Reference: 308

74. Two(2) type of corner wall bracing are:
1. ____________________
2. ____________________

Answer: 1. Let-in bracing
2. Structural sheathing

Question Type: F
Random: Y
Difficulty Level: 2
Chapter Number: 12
Page Reference: 308

75. The notch cut in a rafter where it sets on the top plate is a(n) ________________ cut.

Answer: bird's-mouth

Question Type: F
Random: Y
Difficulty Level: 2
Chapter Number: 12
Page Reference: 312

76. _________________________ are short members place 1/3 the distance down from the ridge to stiff a roof.

Answer: Collar beams

Question Type: F
Random: Y
Difficulty Level: 2
Chapter Number: 12
Page Reference: 317

77. ______________ is a common material used for roof decks.

Answer: Plywood

Question Type: F
Random: Y
Difficulty Level: 1
Chapter Number: 12
Page Reference: 318

78. Six (6) materials used for wall sheathing are:
1. ____________________
2. ____________________
3. ____________________
4. ____________________
5. ____________________
6. ____________________

Answer: 1. Wood board
2. Fiberboard
3. Plywood
4. Gypsum
5. Particleboard
6. Foam sheathing (styrofoam or urethane)

Question Type: F
Random: Y
Difficulty Level: 2
Chapter Number: 12
Page Reference: 319

79. Describe the difference between a "Fink" and "Scissors" truss.

Answer: The main difference between a "Fink" and "Scissors" truss, is the scissors truss has a sloping bottom cord for a vaulted ceiling.

Question Type: E
Random: Y
Difficulty Level: 2
Chapter Number: 12
Page Reference: 327

80. Describe a flitch plate beam.

Answer: A flitch plate beam is where there is a steel plate wit a 2x member on each side, the members are bolted together.

Question Type: S

Random: Y
Difficulty Level: 2
Chapter Number: 12
Page Reference: 330

81. _______________ are used at bottom of masonry walls to let moisture escape to the outside from any interior cavity.

Answer: Weep holes

Question Type: F
Random: Y
Difficulty Level: 2
Chapter Number: 12
Page Reference: 335

82. A masonry wall constructed with a 2" air space between the inner and outer masonry is a(n) __________________ wall.

Answer: cavity

Question Type: F
Random: Y
Difficulty Level: 2
Chapter Number: 12
Page Reference: 338

83. _______________ are used for a fireproof flat roof.

Answer: Concrete joist

Question Type: F
Random: Y
Difficulty Level: 2
Chapter Number: 12
Page Reference: 340

84. The shape of metal studs are ___________________ in section.

Answer: C-shaped

Question Type: F
Random: Y
Difficulty Level: 2
Chapter Number: 12
Page Reference: 340

85. Metal studs are placed in a(n) _________________, which takes the place of a sole plate in wood framing.

Answer: track

Question Type: F
Random: Y
Difficulty Level: 2
Chapter Number: 12
Page Reference: 340

86. Metal framing uses __________________ instead of nails to fasten members together.

Answer: screws

Question Type: F
Random: Y
Difficulty Level: 2
Chapter Number: 12
Page Reference: 340

87. Metal framing uses __________________ to strengthen members at bearing points.

Answer: web stiffeners

Question Type: F
Random: Y
Difficulty Level: 2
Chapter Number: 12
Page Reference: 340

88. Discuss Insulating Concrete Form (ICF) wall systems.

Answer: Insulating Concrete Form (ICF) wall systems are an alternative wall system. They are hollow polystyrene foam blocks, panels, or planks that are erected and then filled with concrete.

Question Type: E
Random: Y
Difficulty Level: 3
Chapter Number: 12
Page Reference: 345

89. Straw bale wall construction is an alternative to wood frame wall construction. True or False?

Answer: True

Question Type: T
Random: Y
Difficulty Level: 1
Chapter Number: 12
Page Reference: 348

90. A roof style that consist of two inclined planes that meet at a ridge over the center of the house and slope down over the side walls is a gable roof. True or False?

Answer: True

Question Type: T
Random: Y
Difficulty Level: 1
Chapter Number: 12
Page Reference: 355

91. Sketch an isometric of a hip roof.

Answer: (see example page 358)

Question Type: s
Random: Y
Difficulty Level: 2
Chapter Number: 12
Page Reference: 358

92. Sketch an isometric of a gambrel roof.

Answer: (see example page 358)

Question Type: s
Random: Y
Difficulty Level: 2
Chapter Number: 12
Page Reference: 358

93. Slate roofing is comes from natural quarry beds. True or False?

Answer: False

Question Type: T
Random: Y

Difficulty Level: 1
Chapter Number: 12
Page Reference: 366

94. The five(5)principal metals used for wall roofing are:
   1. ____________________
   2. ____________________
   3. ____________________
   4. ____________________
   5. ____________________

Answer: 1. Copper
   2. Terneplate
   3. Aluminum
   4. Painted steel
   5. Galvanized iron

Question Type: F
Random: Y
Difficulty Level: 2
Chapter Number: 12
Page Reference: 371

95. Three(3) spray-type insulators are:
   1. ____________________
   2. ____________________
   3. ____________________

Answer: 1. spray cellulose
   2. Two-component isocyanurate foam
   3. Polyurethane

Question Type: F
Random: Y
Difficulty Level: 2
Chapter Number: 12
Page Reference: 383

96. Radiant barriers give off heat to reduce heat loss in a building. True or False?

Answer: False

Question Type: T
Random: Y
Difficulty Level: 1
Chapter Number: 12
Page Reference: 383

97. ______________ is the temperature when air gives off moisture.

Answer: Due point

Question Type: F
Random: Y
Difficulty Level: 2
Chapter Number: 12
Page Reference: 384

98. Moisture moves from place to place in one of four(4) ways:
1. ____________________
2. ____________________
3. ____________________
4. ____________________

Answer: 1. Gravity
2. Capillary action
3. Air flow
4. Diffusion

Question Type: F
Random: Y
Difficulty Level: 2
Chapter Number: 12
Page Reference: 384

99. ______________ are used to reduce the outward migration of moisture-filled air.

Answer: Air barriers

Question Type: F
Random: Y
Difficulty Level: 2
Chapter Number: 12
Page Reference: 385

100. One type of Vapor Diffusion Retarders is VDR paint. True or False?

Answer: True

Question Type: T
Random: Y
Difficulty Level: 2
Chapter Number: 12

Page Reference: 387

MULTIPLE CHOICE: Circle the letter indicating the best choice for each question.

101. Where should the wire mesh reinforcing be placed in a concrete slab-on-ground?
    a. Top of the slab
    b. Center of the slab
    c. Bottom of the slab

Answer: a

Question Type: M
Random: Y
Difficulty Level: 1
Chapter Number: 281

102. Where should the wire mesh reinforcing be placed in a concrete slab-on-ground?
    a. Top of the slab
    b. Center of the slab
    c. Bottom of the slab

Answer: a

Question Type: M
Random: Y
Difficulty Level: 1
Chapter Number: 281

103. In addition to structural design, foundation design should include ________________.

    a. Termite control
    b. Radon control
    c. Moisture control
    d. Insulation (heat loss and heat gain)
    e. All the above

Answer: e

Question Type: M
Random: Y
Difficulty Level: 2
Chapter Number: 12

Page Reference: 259

104. Crawl space ventilation should be the total crawl space area divided by ______________________.
   a. 50
   b. 100
   c. 150
   d. 200
   e. 250

Answer: c

Question Type: M
Random: Y
Difficulty Level: 2
Chapter Number: 12
Page Reference: 272

105. Welded Wire Mesh with a designation of 6 x 6 w 1.4 x w 1.4, the wire spacing is ______________________.
   a. 6"
   b. 36"
   c. 1.4"
   d. 1.4 mm
   e. All the above

Answer: a

Question Type: M
Random: Y
Difficulty Level: 2
Chapter Number: 12
Page Reference: 274

106. A good rule of thumb is to limit the height of a unreinforced concrete foundation to _________ times the thickness of the wall.
   a. 5
   b. 10
   c. 20
   d. 50
   e. 100

Answer: b

Question Type: M
Random: Y
Difficulty Level: 2
Chapter Number: 12

Page Reference: 277

107. Wall studs in platform construction set on a(n)_______________.
   a. subfloor
   b. sill plate
   c. sole plate
   d. header
   e. band board

Answer: c

Question Type: M
Random: Y
Difficulty Level: 2
Chapter Number: 12
Page Reference: 305

108. In roofing terminology, a square is an area of ___________ sqft.
   a. 10
   b. 25
   c. 50
   d. 75
   e. 100

Answer: e

Question Type: M
Random: Y
Difficulty Level: 2
Chapter Number: 12
Page Reference: 361

Chapter 13. STRUCTURAL MEMBER SELECTION

1. What lumber grades are generally used for the major amount of residential framing throughout the country?

Answer: Select structural No.1, No.2, No.3, as well as appearance and stud grade lumber is generally used for the major amount of residential framing throughout the country.

Question Type: S
Random: Y
Difficulty Level: 2
Chapter Number: 13
Page Reference: 398

2. What physical characteristics must structural lumber possess?

Answer: Structural lumber should have few physical defects, and have a large modules of elasticity which produces stiffness and resistance to deflection.

Question Type: S
Random: Y
Difficulty Level: 2
Chapter Number: 13
Page Reference: 398

3. Explain the word deflection as used in structural terminology.

Answer: Deflection in structural terminology describes woods tendencies to move or sag when put under stress such as supporting a load.

Question Type: S
Random: Y
Difficulty Level: 2
Chapter Number: 13
Page Reference: 398

4. Wood columns used to support a girder in a structure would be subjected to what type of stresses?

Answer: Wood columns used to support a girder in a structure would be subjected to the forces of compression.

Question Type: S
Random: Y
Difficulty Level: 2

Chapter Number: 13
Page Reference: 399

5. Local building codes require what minimum live load on living room floors?

Answer: Local building codes require a minimum live load on living room floors of 40 psf.

Question Type: S
Random: Y
Difficulty Level: 1
Chapter Number: 13
Page Reference: 400

6. Why must both the live load and the dead load be considered when determining the size of a structural framing member?

Answer: Both the live load and the dead load must be considered when determining the size of a structural framing member for most structural framing lumber will be subjected to both dead loads and live loads.

Question Type: S
Random: Y
Difficulty Level: 2
Chapter Number: 13
Page Reference: 400

7. Define the term safety factor.

Answer: All structural framing charts list the maximum allowable loads that a structural member can safely support. Codes require that a safety factor be included in the allowable span tables. Designers who have doubts about a member'- capability should up size the member.

Question Type: S
Random: Y
Difficulty Level: 2
Chapter Number: 13
Page Reference: 397

8. Differentiate between light roofing and heavy roofing when sizing rafters.

Answer: When sizing rafters, one must differentiate between light roofing and heavy roofing materials. Fiberglass or wood shingles are considered light roofing materials. elate or mission tile are considered heavy roofing materials.

Question Type: S
Random: Y
Difficulty Level: 1
Chapter Number: 13
Page Reference: 415

9. Why is it important in conventional roof framing to have ceiling joists parallel to the rafters?

Answer: Ceiling joists need to run parallel to the rafters in conventional roof framing to counter act the forces of lateral thrust. The ceiling joists help to stabilize the walls and rafters.

Question Type: S
Random: Y
Difficulty Level: 1
Chapter Number: 13
Page Reference: 413

10. In planning a plank and beam roof with wide spacing of the beams, why is the stiffness of the planking an important consideration?

Answer: The planking provides lateral stiffness to a plank and beam roof with wide spacing on the beam.

Question Type: S
Random: Y
Difficulty Level: 2
Chapter Number: 13
Page Reference: 422

11. Is the rafter length or the rafter span used in selecting a proper size for a rafter?

Answer: The rafter span (horizontal distance not the diagonal distance) is used in selecting a proper size rafter.

Question Type: S
Random: Y
Difficulty Level: 1
Chapter Number: 13
Page Reference: 413

12. What are the two (2) most common species of wood used in structural framing?

1. ____________________________

2. ____________________________

Answer: The two most common species of wood used in structural framing are southern pine and douglas fir.

Question Type: S
Random: Y
Difficulty Level: 1
Chapter Number: 13
Page Reference: 399

13. What are the five (5) stresses acting upon structural members?

1. ____________________________

2. ____________________________

3. ____________________________

4. ____________________________

5. ____________________________

Answer: 1. Compression 2. Tension 3. Bending 4. Shear 5. Buckling

Question Type: F
Random: Y
Difficulty Level: 2
Chapter Number: 13
Page Reference: 399

14. What is "deflection"?

Answer: Deflection is the tendency for a member to bend under loads.

Question Type: S
Random: Y
Difficulty Level: 1
Chapter Number: 13
Page Reference: 398

15. What is "creep" and how is it overcome?

Answer: Creep is long-term deflection caused by continuous heavy loads on a member over a long period of time. To overcome creep, stiffer and larger members must be used.

Question Type: S
Random: Y
Difficulty Level: 1
Chapter Number: 13
Page Reference: 399

15. What is "elasticity" (E)?

Answer: Elasticity (E) is the measurement of stiffness in a member. The greater the "modules of elasticity" the more stiff the member.

Question Type: S
Random: Y
Difficulty Level: 2
Chapter Number: 13
Page Reference: 399

16. The only stress to be considered in column design is compression. True or False?

Answer: False

Question Type: T
Random: Y
Difficulty Level: 2
Chapter Number: 13
Page Reference: 399

17. What is "shear" in a member?

Answer: Shear is a sliding or cutting stress as the result of a vertical load applied to a member which causes the member to fail near its support.

Question Type: S
Random: Y
Difficulty Level: 2
Chapter Number: 13
Page Reference: 399

18. What is the maximum percentage of moisture allowed in seasoned residential framing lumber and why?

Answer: The maximum percentage of moisture allowed in residential framing lumber is 19%. When moisture is removed from the wood, the wood shrinks causing the wood to become more dense and stronger. Any wood with a greater moisture content than 19% is considered unseasoned. Some structural members have a 15% limit of moisture.

Question Type: S
Random: Y
Difficulty Level: 2
Chapter Number: 13
Page Reference: 399

19. Finish lumber is usually specified with a maximum percentage of moisture of ________________ percent.

Answer: Twelve (12)

Question Type: F
Random: Y
Difficulty Level: 1
Chapter Number: 13
Page Reference: 400

20. Define the two (2) major loads that are applied to a structural member.

Answer: The two major loads applied to a structural member are the dead load and the live load. The dead load consists of the weight of the structure and the permanent equipment. The live load consists of the people, furniture and other non-stationary items.

Question Type: E
Random: Y
Difficulty Level: 2
Chapter Number: 13
Page Reference: 400

21. Beams or girders that support first floor joists are indicated on the ______________________ or ______________________ plan.

Answer: Foundation or basement plan.

Question Type: F
Random: Y
Difficulty Level: 1
Chapter Number: 13
Page Reference: 401

22. Define a LVL member.

Answer: A (LVL), is a Laminated Veneer Lumber member.

Question Type: S
Random: Y
Difficulty Level: 2
Chapter Number: 13
Page Reference: 402

23. Define a PSL member.

Answer: A (PSL), is a Parallel-Strand Lumber member.

Question Type: s
Random: Y
Difficulty Level: 2
Chapter Number: 13
Page Reference: 402

24. In conventional roof framing how is the lateral thrust of the rafter counteracted?

Answer: The lateral thrust of the rafter is counteracted by running the ceiling joists parallel to the rafters and fastening the rafter and the ceiling joist together.

Question Type: S
Random: Y
Difficulty Level: 2
Chapter Number: 13
Page Reference: 411

25. The span of a rafter is the length of the slope of the rafter from bearing to ridge board. True or False?

Answer: False

Question Type: T
Random: Y
Difficulty Level: 2

Chapter Number: 13
Page Reference:

26. What will be the design length of a rafter with a roof slope of 5/12 and a horizontal span of 15'-0"?

Answer: The design length of the rafter will be 15'-0"

Question Type: S
Random: Y
Difficulty Level: 3
Chapter Number: 13
Page Reference:

27. Common spacing for floor joist is ____, _____, and ____ inches oc.

Answer: a. 12"   b. 16"   c. 24"

Question Type: F
Random: Y
Difficulty Level: 3
Chapter Number: 13
Page Reference: 405

28. Explain the meaning of a deflection of L/360.

Answer: The standard maximum allowable deflection for this member would be its length divided 360. A member that spans 15'-0" or 180"(L)/360 = 1/2" maximum deflection.

Question Type: S
Random: Y
Difficulty Level: 3
Chapter Number: 13
Page Reference: 405

29. The two common allowable deflections are ________ and _________.

Answer: L/240 and 1/360

Question Type: F
Random: Y
Difficulty Level: 3
Chapter Number: 13
Page Reference: 405

MULTIPLE CHOICE: Circle the letter indicating the best choice for each question.

30. What is the minimum live load for bedrooms allowed by the building codes?
    a. 20 p.s.f.
    b. 30 p.s.f.
    c. 40 p.s.f.
    d. 60 p.s.f.
    e. 100 p.s.f.

Answer: b

Question Type: M
Random: Y
Difficulty Level: 1
Chapter Number: 13
Page Reference: 400

Chapter 14. ARCHITECTURAL DETAILS

1. What factors are used to determine the choice of scale when drawing details?

Answer: Choosing scale depends upon the size of the members in the detail, the amount of detail needed, and the space available on the layout sheet.

Question Type: S
Random: Y
Difficulty Level: 2
Chapter Number: 14
Page Reference: 433

2. What information does a typical wall detail usually show?

Answer: A typical wall detail should incorporate the footing detail, sill detail, wall detail, and roof connection detail; all should be properly in line for relative reference.

Question Type: S
Random: Y
Difficulty Level: 2
Chapter Number: 14
Page Reference: 433

3. Why are structural members in detail drawings drawn actual size rather than nominal size?

Answer: Structural members must be drawn to their actual size in details since the details illustrate the precise way the materials are to be assembled during construction.

Question Type: S
Random: Y
Difficulty Level: 2
Chapter Number: 14
Page Reference: 434

4. In gable-roof terminology, what is the difference between "span" and "run"?

Answer: In gable-roof terminology the span is the distance between supports and run is one half that distance.

Question Type: S

Random: Y
Difficulty Level: 2
Chapter Number: 14
Page Reference: 441

5. Why must manufacturers' catalogues be consulted when determining rough-opening sizes for windows?

Answer: Manufacturers' catalogues contain detailed information on rough-openings for windows. These openings may vary depending on the type of building materials used as well a. methods of attaching the windows in the openings.

Question Type: E
Random: Y
Difficulty Level: 2
Chapter Number: 14
Page Reference: 446

6. Define a window sash.

Answer: The window sash is the part of the window that usually moves inside a frame. The sash is the glass and the members surrounding the glass. The sash of a double hung window moves up and down. The sash of a casement window opens to the exterior. Some windows have a fixed sash that does not move.

Question Type: S
Random: Y
Difficulty Level: 2
Chapter Number: 14
Page Reference: 446

7. What dimensions are first selected in fireplace design, and what should be the basis for their selection?

Answer: The first dimensions selected in fireplace design are the openings. Openings should not only have correct proportions to insure proper functioning but also have a pleasing proportion in relation to the size of the room. Rectangles appear more desirable than squares.

Question Type: E
Random: Y
Difficulty Level: 2
Chapter Number: 14
Page Reference: 454

8. For a workable fireplace, what should be the proportion of the cross-sectional flue area to the fireplace opening area?

Answer: Fireplaces should have a flue area one-tenth the size of the fireplace opening area. Chimney heights under 14' may require even larger flue sizes.

Question Type: S
Random: Y
Difficulty Level: 2
Chapter Number: 14
Page Reference: 454

9. What two dimensions are the most critical in satisfactory stair design?

Answer: The most important dimensions in the stairs are the risers and treads.

Question Type: S
Random: Y
Difficulty Level: 2
Chapter Number: 14
Page Reference: 467

10. Describe how an architect's scale can be used to lay out the riser height and tread width of stairs in elevation or section without measuring each riser and tread from floor to floor.

Answer: The best way to lay out stairs on a drawing is with an architects scale. Establish the floor to floor height and draw floor lines between which the stair is to be drawn. Divide the distance (in inches) between floor. by seven to determine the number of risers needed. If 16 risers are required, place the same scale used for the layout with 0 (zero) on the scale located on the top floor and place 16 on the scale on the lower floor and mark off the risers. If 16 risers are required, the stair will need 15 treads. Measure 11" for the first tread at the top of the stair. Place a scale with the 15th increment on the first floor level directly below the top riser on the second floor. Incline the scale until the lath increment aligns with the vertical line that you drew to locate the first 11" tread at the top of the stair. Mark off each full number increment on the scale to locate the treads.

Question Type: E
Random: Y

Difficulty Level: 3
Chapter Number: 14
Page Reference: 468

11. Write the three formulas for checking riser and tread proportions for an interior residential stairs.
    1. ____________________
    2. ____________________
    3. ____________________

Answer: There are three formulas for figuring riser and tread:
1. 2 risers + 1 tread = between 24" and 25"
2. Riser times tread = between 72" and 77"
3. Riser + tread = between 17" and 18"

Question Type: F
Random: Y
Difficulty Level: 2
Chapter Number: 14
Page Reference: 464

12. What is the fire door and what do the labels A, B, and C designate?

Answer: A fire rated door is a door that will withstand a fire for a designated time before disintegrating. Fire doors have class ratings or labels showing how long they will last during a fire. A "C" rated door will withstand fire for 45 minutes; and "B" rated doors will withstand fire for 1 hour and 1 1/2 hours; and an "A" rated door will withstand fire for 3 hours.

Question Type: S
Random: Y
Difficulty Level: 2
Chapter Number: 14
Page Reference: 450

13. What are anchor bolts and where are they used in wood frame construction?

Answer: Anchor bolts are steel bolts placed in foundation walls and used to hold sill plates to the foundation.

Question Type: S
Random: Y
Difficulty Level: 2
Chapter Number: 14

Page Reference: 435-6

14. What is used to support brick masonry above a door or window opening?

Answer: Steel angles (lintels) are used to support brick over windows and doors.

Question Type: S
Random: Y
Difficulty Level: 2
Chapter Number: 14
Page Reference: 438

15. Residential doors are usually 6'-8" high. What is the common height of commercial doors?

Answer: The common height for a commercial door is 7'-0".

Question Type: S
Random: Y
Difficulty Level: 1
Chapter Number: 14
Page Reference: 450

16. What is the minimum headroom above stairs?

Answer: The minimum headroom above stairs is 6'-8".

Question Type: S
Random: Y
Difficulty Level: 2
Chapter Number: 14
Page Reference: 466

17. Where is a brick rowlock used in window detailing?.

Answer: A brick rowlock is used as a window sill.

Question Type: S
Random: Y
Difficulty Level: 2
Chapter Number: 14
Page Reference: 439

18. What is a "bird's-mouth"?

Answer: A bird's mouth is a notch on a rafter forming a flat horizontal bearing surface for the rafter. The surface of the bird 's-mouth should be a minimum of three inches (3") wide since the notch supports the main weight of the roof.

Question Type: S
Random: Y
Difficulty Level: 2
Chapter Number: 14
Page Reference: 435-6

19. Structural members in details should be drawn at nominal size rather than actual size. True or False?

Answer: False

Question Type: T
Random: Y
Difficulty Level: 2
Chapter Number: 14
Page Reference: 434

20. What are the three (3) methods of indicating roof slopes or pitches on drawings?
1. ____________________
2. ____________________
3. ____________________

Answer: 1. A slope-ratio diagram 2. A fractional pitch indication 3. An angular dimension

Question Type: F
Random: Y
Difficulty Level: 2
Chapter Number: 14
Page Reference: 439

21. As a rule of thumb, the total window area of a room should be ____________ percent of the total floor space of the room.

Answer: Ten percent.

Question Type: F
Random: Y
Difficulty Level: 1
Chapter Number: 14

Page Reference: 442

22. Name five (5) different types of windows.
    1. ______________________________
    2. ______________________________
    3. ______________________________
    4. ______________________________
    5. ______________________________

Answer: 1. Double hung
        2. Single hung
        3. Sliding
        4. Fixed
        5. Casement

Question Type: F
Random: Y
Difficulty Level: 1
Chapter Number: 14
Page Reference: 444-6

23. Describe the difference between a "hopper" and an "awning" window.

Answer: A hopper window has a small sash, opening either in or out, and is usually combined with larger fixed window. An awning window opens out horizontally with the hinges of each sash at the top.

Question Type: S
Random: Y
Difficulty Level: 2
Chapter Number: 14
Page Reference: 444-6

24. Describe a clerestory window.

Answer: A clerestory window is a window or series of windows high in a wall placed above a lower roof level. Their height affords privacy and permits them to cast dramatic light effects in a room.

Question Type: S
Random: Y
Difficulty Level: 2
Chapter Number: 14
Page Reference: 446

25. What do the "rough opening" dimensions of a window indicate?

Answer: The rough opening dimensions of a window indicate the width and height of the framed opening that receives the window.

Question Type: S
Random: Y
Difficulty Level: 3
Chapter Number: 14
Page Reference: 446

26. What are the basic steps taken in drawing a window detail?
1. ______________________________
2. ______________________________
3. ______________________________
4. ______________________________

Answer: The basic steps taken in drawing a window detail are:
1. Establish the rough opening of the masonry opening or the wood framing members.
2. Draw the sheathing and interior wall construction.
3. Locate the window frame in relation to the rough opening.
4. Complete the sash, trim and exterior wall covering.

Question Type: F
Random: Y
Difficulty Level: 3
Chapter Number: 14
Page Reference: 448

27. Describe a "hollow core" wood flush door.

Answer: A hollow core wood flush door is built of solid wood stiles and rails with various wood veneering glued to both sides, producing smooth, easy to paint or stain surfaces. The interior of a hollow core door has grillage-filled cavities, making it lighter and ideal for interior use.

Question Type: E
Random: Y
Difficulty Level: 2
Chapter Number: 14
Page Reference: 449

28. Describe a "solid core" wood flush door.

Answer: A solid core wood flush door is built of solid wood stiles and rails with various wood veneering glued to both sides, producing smooth, easy to paint or stain surfaces. The interior of a solid core door has the cavity filled with wood blocking or other dense material making it heavier and more appropriate for exterior entrances.

Question Type: E
Random: Y
Difficulty Level: 2
Chapter Number: 14
Page Reference: 449

29. Describe a wood panel door.

Answer: A wood panel door is usually constructed of ponderosa pine or douglas fir and consists of various arrangements of panels and lights surrounded with stiles and rails, usually of solid wood, to provide different appearances. Some panel doors are manufactured with glue-and-veneer stiles and rails.

Question Type: E
Random: Y
Difficulty Level: 2
Chapter Number: 14
Page Reference: 450

30. Name four (4) variations of a wood panel door.

1. ______________________________
2. ______________________________
3. ______________________________
4. ______________________________

Answer: 1. French door 2. Dutch door 3. Louver door 4. Jalousie door

Question Type: F
Random: Y
Difficulty Level: 2
Chapter Number: 14
Page Reference: 450

31. Sliding doors are manufactured of aluminum only. True or False?

Answer: False

Question Type:
Random: Y
Difficulty Level:
Chapter Number: 14
Page Reference: 453

32. What is the difference between a wood solid core door and a wood fire rated door?

Answer: A wood solid core door is filled with solid wood material while a wood fire rated door is filled with a fire-resisting material and treated with fire-resisting chemicals.

Question Type: S
Random: Y
Difficulty Level: 2
Chapter Number: 14
Page Reference: 450

33. In masonry construction, what is used to support the masonry above the door opening?

Answer: A steel or pre-cast lintel must be used to support the masonry above a door opening.

Question Type: S
Random: Y
Difficulty Level: 2
Chapter Number: 14
Page Reference: 453

34. Name three (3) types of conventional single-face fireplaces.
1. ____________________
2. ____________________
3. ____________________

Answer: 1. Wall 2. Corner 3. Back-to-back

Question Type: F
Random: Y
Difficulty Level: 2
Chapter Number: 14
Page Reference: 454

35. What are the six (6) critical points that must be considered in single face fireplace design and construction?
1. ____________________
2. ____________________
3. ____________________
4. ____________________
5. ____________________
6. ____________________

Answer: 1. Size of the fireplace opening 2. Flue size and chimney height 3. Shape of the combustion chamber 4. Design of the throat 5. Shape of the smoke shelf and smoke dome 6. Support for the hearth.

Question Type: F
Random: Y
Difficulty Level: 2
Chapter Number: 14
Page Reference: 454

36. How far above the high point of the roof should a chimney extend?

Answer: A chimney should extend a minimum of two feet (2'-0") above the high point of the roof.

Question Type: S
Random: Y
Difficulty Level: 2
Chapter Number: 14
Page Reference: 459

37. What is the material that lines the combustion chamber of a solid masonry fireplace?

Answer: The material that lines the combustion chamber of a fireplace is firebrick manufactured from a heat resistant clay, it is 4" thick and is set in a fire clay mortar.

Question Type: S
Random: Y
Difficulty Level: 2
Chapter Number: 14
Page Reference: 454

38. What does the term "headroom" in stair construction indicate?

Answer: The term headroom indicates the narrowest distance between the surface of a tread and any ceiling or header above.

Question Type: S
Random: Y
Difficulty Level: 2
Chapter Number: 14
Page Reference: 466

39. Define a balusters.

Answer: Thin vertical supports for the handrail of open stairs.

Question Type: S
Random: Y
Difficulty Level: 2
Chapter Number: 14
Page Reference: 461

40. Define a newel post.

Answer: Main post of the railing at the bottom of the stairs.

Question Type: S
Random: Y
Difficulty Level: 2
Chapter Number: 14
Page Reference: 461

41. Define rise.

Answer: The total floor-to-floor vertical height of a stair.

Question Type: S
Random: Y
Difficulty Level: 2
Chapter Number: 14
Page Reference: 461

42. Define a riser.

Answer: The vertical face of the step.

Question Type: S
Random: Y
Difficulty Level: 2
Chapter Number: 14
Page Reference: 461

43. Define a tread.

Answer: The horizontal surface member of each step.

Question Type: S
Random: Y
Difficulty Level: 2
Chapter Number: 14
Page Reference: 461

44. Define a winder.

Answer: The radiating or wedge-shaped treads at turns of stairs.

Question Type: S
Random: Y
Difficulty Level: 2
Chapter Number: 14
Page Reference: 461

45. What is the most comfortable handrail height and how is it measured?

Answer: The most comfortable handrail height is 32" and is measured vertically from the front edge of the tread to the top of the handrail.

Question Type: s
Random: Y
Difficulty Level: 2
Chapter Number: 14
Page Reference: 468

46. Railings should be ___________ high on residential landings and _______________ high on commercial landings.

Answer: 36" and 42" respectfully.

Question Type: F
Random: Y
Difficulty Level: 2
Chapter Number: 14
Page Reference: 466

47. The National Fenestration Rating Council (NFRC) uses the _________ value to rate the energy performance of windows.

Answer: U-value

Question Type: F
Random: Y
Difficulty Level: 2
Chapter Number: 14
Page Reference: 441

48. When choosing a window, the three (3) things to consider are:
1. _______________

2. ____________________
3. ____________________

Answer: 1.Type of window 2. Type of glass 3. Type of material used for frame.

Question Type: F
Random: Y
Difficulty Level: 2
Chapter Number: 14
Page Reference: 442

49. Describe a palladian window.

Answer: A palladian window has a tall semicircular topped center section with straight-headed side windows on each side.

Question Type: S
Random: Y
Difficulty Level: 2
Chapter Number: 14
Page Reference: 446

50. Define the meaning of.

Answer: The "rough opening" is the frame opening in a wall to receive a window. This opening is usually 1/4" larger per side than the windows unit dimension.

Question Type: S
Random: Y
Difficulty Level: 2
Chapter Number: 14
Page Reference: 446

51. Horizontal members of a door are ______________________________.

Answer: Horizontal members of a door are rails.

Question Type: F
Random: Y
Difficulty Level: 2
Chapter Number: 14
Page Reference: 449

52. Vertical members of a door are __________________________.

Answer: Vertical members of a door are stiles.

Question Type: F
Random: Y
Difficulty Level: 2
Chapter Number: 14
Page Reference: 449

53. Define nosing.

Answer: The round projection of the tread beyond the face of the riser.

Question Type: S
Random: Y
Difficulty Level: 2
Chapter Number: 14
Page Reference: 461

MULTIPLE CHOICE: Circle the letter indicating the best choice for each question.

54. The width of the air space provided between the masonry chimney and wood framing is:
    a. 1"
    b. 1 1/2.
    c. 2"
    d. 2 1/2"
    e. 3"

Answer: c

Question Type: M
Random: Y
Difficulty Level: 2
Chapter Number: 14
Page Reference: 457

55. Exterior wood door frames are usually:
    a. 3/4" thick
    b. 1" thick
    c. 1 1/16" thick
    d. 1 1/4" thick
    e. 1 1/2" thick

Answer: c

Question Type: M
Random: Y
Difficulty Level: 2
Chapter Number: 14
Page Reference: xx

Chapter 15. BASIC RESIDENTIAL PLANNING

1. Custom home designs vary, therefore a planning checklist is not important. True or False?

Answer: False

Question Type: T
Random: Y
Difficulty Level: 1
Chapter Number: 15
Page Reference: 502

2. Do contour lines crossing a small stream point upstream or downstream?

Answer: Contour lines crossing a small stream point upstream.

Question Type: S
Random: Y
Difficulty Level: 1
Chapter Number: 15
Page Reference: 506

3. Why is a southern exposure generally the most satisfactory direction in solar orienting a window wall?

Answer: A southern exposure is generally the most satisfactory direction in solar orienting a window wall, because with correctly designed overhangs and sunshades, window walls can restrict the sun in summer and provide warmth and light within the house in the winter.

Question Type: S
Random: Y
Difficulty Level: 2
Chapter Number: 15
Page Reference: 512

4. Which part of a home should have the most isolation and quiet?

Answer: Bedrooms, libraries and private spaces should have the most isolation and quiet.

Question Type: S
Random: Y
Difficulty Level: 1

Chapter Number: 15
Page Reference: 513

5. What geometric shape is the most economical for a basic plan? Explain

Answer: The square shape is the most economical for a basic plan because it has the least exterior surface which is the most expensive to construct.

Question Type: E
Random: Y
Difficulty Level: 3
Chapter Number: 15
Page Reference:

6. Evaluate the traffic flow in our classroom from the standpoint of convenience.

Answer: (Answers will vary)

Question Type: E
Random: Y
Difficulty Level: 2
Chapter Number: 15
Page Reference:

7. The maximum length of a satisfactory kitchen work triangle is 10 feet. True or False?

Answer: False

Question Type: T
Random: Y
Difficulty Level: 1
Chapter Number: 15
Page Reference: 535

8. Define fenestration.

Answer: Fenestration is the placement and sizing of window and door openings.

Question Type: S
Random: Y
Difficulty Level: 2
Chapter Number: 15

Page Reference: 540

9. Why are simple, uncomplicated exteriors usually the most pleasing?

Answer: Simple, uncomplicated exteriors are usually the most pleasing because the designer can create unity, harmony, scale, balance, emphasis, and focal point. Complicated exteriors often have competing elements that de-emphasize important features and overpower the focal point.

Question Type: E
Random: Y
Difficulty Level: 2
Chapter Number: 15
Page Reference: 538

10. What is the purpose of a planning checklist?

Answer: Since almost every planning project brings together dissimilar combinations of tastes and needs, planning often becomes a matter of compromise. To be certain that important features are not overlooked or forgotten, a procedure must be followed which will organize these requirements before planning is started. This procedure is known as a checklist.

Question Type: S
Random: Y
Difficulty Level: 1
Chapter Number: 15
Page Reference: 501

11. A residential planning checklist contains seven major headings. Name any five of these headings.

1. ____________________

2. ____________________

3. ____________________

4. ____________________

5. ____________________

Answer: 1. Site 2. Occupants 3. Individual requirements 4. General design 5. Budget restrictions 6. Mechanical equipment required 7. Storage areas required

Question Type: F
Random: Y
Difficulty Level: 1
Chapter Number: 15
Page Reference: 502

12. What is the first thing a planner must do when starting to design a project?

Answer: The first thing a planner must do when starting a project is make a thorough investigation of the property to determine the restrictions placed upon the property by zoning regulations and ordinances, easements, right-of-ways, etc.

Question Type: E
Random: Y
Difficulty Level: 2
Chapter Number: 15
Page Reference: 503

13. What are the owner's responsibilities during the planning process?

Answer: The responsibility of the owner during the planning period is to select the property and to have the clear-title legality of the deed to the property investigated.

Question Type: S
Random: Y
Difficulty Level: 2
Chapter Number: 15
Page Reference: 503

14. What is a "plat"?

Answer: A plat is a graphic description of the property.

Question Type: S
Random: Y
Difficulty Level: 1
Chapter Number: 15
Page Reference: 503

15. What constitutes the legal documents of the property?

Answer: The plat and a written legal description of the property constitute the legal documents.

Question Type: S
Random: Y
Difficulty Level: 1
Chapter Number: 15
Page Reference: 503

16. A plat and a site plan give the same information and are inter-changeable. True or False?

Answer: False

Question Type: T
Random: Y
Difficulty Level: 2
Chapter Number: 15
Page Reference: 503

17. What is a contour line and what is its purpose?

Answer: A contour line is a symbol indicating elevation or level variations of the property. Each contour is plotted to represent similar elevations along the entire line. Its purpose is to indicate the various levels and slopes of the property.

Question Type: S
Random: Y
Difficulty Level: 1
Chapter Number: 15
Page Reference: 503

18. What is a "contour interval"?

Answer: A contour interval is the vertical distance between contour levels.

Question Type: S
Random: Y
Difficulty Level: 1
Chapter Number: 15
Page Reference: 504

19. What information does a "site analysis" contain?

Answer: A site analysis contains information on existing and future conditions which may impact upon the site and proposed building.

Question Type: S
Random: Y
Difficulty Level: 1
Chapter Number: 15
Page Reference: 507

20. There are seven (7) major topics to be investigated for a site analysis. Name these seven major topics.

1. ____________________

2. ____________________

3. ____________________

4. ____________________

5. ____________________

6. ____________________

7. ____________________

Answer: 1. Climate 2. Circulation 3. Neighborhood 4. Natural elements 5. Man-made additions 6. Sensory features 7. Zoning

Question Type: F
Random: Y
Difficulty Level: 2
Chapter Number: 15
Page Reference: 507

21. Man-made additions on a site include which of the following?
a. Fences
b. Retaining walls
c. Pavements and streets
d. Sidewalks
e. All of the above

Answer: e

Question Type: M
Random: Y
Difficulty Level: 1

Chapter Number: 15
Page Reference: 507

22. What is meant by an "easement"?

Answer: An easement is a the legal means by which a party other than the owner has been given access or certain rights to the property for various reasons. The right is irrevocable and must be honored by the owner and future owners.

Question Type: S
Random: Y
Difficulty Level: 2
Chapter Number: 15
Page Reference: 503

23. What is the advantage of orienting the residence to the rear of the property?

Answer: Orienting the residence to the rear of the property utilizes a large portion of the property for livability, includes inexpensive, exterior level areas into the overall planning making the residence appear more spacious and provides privacy from noisy public streets.

Question Type: E
Random: Y
Difficulty Level: 2
Chapter Number: 15
Page Reference:

24. Why should deciduous trees be placed on the southern side of the building?

Answer: Deciduous trees should be placed on the southern side of the building because they afford shade during the summer months and lose there leaves during the winter allowing the warmth of the sun to enter the building.

Question Type: S
Random: Y
Difficulty Level: 1
Chapter Number: 15
Page Reference: 515

25. What areas of the residence should be isolated from

noise-producing conditions?

a. Living areas
b. Sleeping areas
c. Recreation areas
d. Utility areas
e. Entertaining areas

Answer: b

Question Type: M
Random: Y
Difficulty Level: 1
Chapter Number: 15
Page Reference: 527

26. ______________________ are good, inexpensive areas of the structure for storage of solar heat.

Answer: Masonry floors or walls

Question Type: F
Random: Y
Difficulty Level: 2
Chapter Number: 15
Page Reference: 517

27. What is the difference between active and passive solar features in a residence?

Answer: Passive solar features have few moving parts and no operating or maintenance costs. The solar heat gain is collected in inert thermal masses and transferred to adjacent areas by natural radiation. Active solar features are mechanical devices that transfer the heat collected from the exterior to various parts of the residence.

Question Type: E
Random: Y
Difficulty Level: 3
Chapter Number: 15
Page Reference: 514

28. Passive solar features must be included in the design of the residence. True or False?

Answer: True

Question Type: T

Random: Y
Difficulty Level: 1
Chapter Number: 15
Page Reference: 514

29. Name three (3) methods of collecting and storing solar heat.

1. ______________________

2. ______________________

3. ______________________

Answer: 1. "Trombe" walls or thermal masses 2. Water filled tanks or tubes 3. Underground rock bins

Question Type: F
Random: Y
Difficulty Level: 1
Chapter Number: 15
Page Reference: 517-18

30. Why is an earth sheltered building a good alternative to solar heat storage?

Answer: An earth sheltered building is a good alternative because the earth maintains a more stable and moderate temperature than the air above requiring less heat in the winter and less cooling in the summer to maintain comfortable temperatures.

Question Type: E
Random: Y
Difficulty Level: 2
Chapter Number: 15
Page Reference: 516

31. Why is the entry to a residence important in the total design?

Answer: The entry is important in the total design because it is a transitional space from the exterior to the interior, it is a space for the removal and storage of outer garments, it provides an orientation to the house for visitors and it sets the architectural character for the residence. Its location and design should be obvious and inviting to both the occupants and guests.

Question Type: E
Random: Y
Difficulty Level: 2
Chapter Number: 15
Page Reference: 541

32. Define the kitchen "work triangle".

Answer: The kitchen work triangle is the space inside the triangle connecting the sink, cooking area and refrigerator.

Question Type: S
Random: Y
Difficulty Level: 2
Chapter Number: 15
Page Reference: 535

33. What is the ideal length of the perimeter of the kitchen "work triangle"?

Answer: The ideal length of the perimeter of the kitchen work triangle is 12 feet to 20 feet maximum.

Question Type: S
Random: Y
Difficulty Level: 1
Chapter Number: 15
Page Reference: 535

34. How can unity and harmony be achieved in residential design?

Answer: Unity and harmony can be achieved in residential design by careful placement of repetitive elements such as windows.

Question Type: S
Random: Y
Difficulty Level: 2
Chapter Number: 15
Page Reference: 540

35. Name the three (3) most important requirements in the selection of windows for a residence.

1. ______________________________

2. ______________________________

3. ______________________________

Answer: 1. Room use   2. Ventilation   3. Site orientation

Question Type: F
Random: Y
Difficulty Level: 1
Chapter Number: 15
Page Reference: 540

36. Fenestration refers to exterior finishes. True or False.

Answer: False.

Question Type: T
Random: Y
Difficulty Level: 2
Chapter Number: 15
Page Reference: 540

37. What are the two styles of doors used for front entrances?

1. ______________________________

2. ______________________________

Answer: 1. Panel with various adaptations   2. Flush with or without lights

Question Type: F
Random: Y
Difficulty Level: 1
Chapter Number: 15
Page Reference: 541

38. Exterior trim details should be consistent with the character of the elevations. True or False?

Answer: True

Question Type: T
Random: Y
Difficulty Level: 2
Chapter Number: 15
Page Reference: 541

39. The four sides of the residence should be compatible in design

and carry out an interesting theme. True or False?

Answer: True

Question Type: T
Random: Y
Difficulty Level: 1
Chapter Number: 15
Page Reference: 541

40. Contour intervals should change on a site plan. True or False?

Answer: False.

Question Type: T
Random: Y
Difficulty Level: 1
Chapter Number: 15
Page Reference: 504

41. With today's excavation equipment, the designer needs not worry about the amount of excavation required with their design. True or False?

Answer: False.

Question Type: T
Random: Y
Difficulty Level: 1
Chapter Number: 15
Page Reference: 511

42. Discuss cluster housing.

Answer: Cluster housing is where houses are placed together with common party walls to conserve land usage. This is only available in new developments.

Question Type: E
Random: Y
Difficulty Level: 2
Chapter Number: 15
Page Reference: 510

43. A(n) _______________ is a convenient aid in developing and understanding the relationships among spaces in a residence.

Answer: Matrix

Question Type: F
Random: Y
Difficulty Level: 2
Chapter Number: 15
Page Reference: 521

44. A common scale to establish a differential in a matrix may range from ______________________ to ______________________.

Answer: Essential to detrimental.

Question Type: F
Random: Y
Difficulty Level: 2
Chapter Number: 15
Page Reference: 521

45. Discuss a "bubble sketch".

Answer: Bubble sketches are freehand circles or bubbles drawn to represent each area or room of a building.

Question Type: E
Random: Y
Difficulty Level: 2
Chapter Number: 15
Page Reference: 522

46. ____________________ is when a design contains fewer, but larger rooms that flow together.

Answer: Open planning.

Question Type: F
Random: Y
Difficulty Level: 2
Chapter Number: 15
Page Reference: 526

47. A designer may conserve floor space by making areas serve dual purposes. True or False?

Answer: True.

Question Type: F

Random: Y
Difficulty Level: 2
Chapter Number: 15
Page Reference: 524

48. In a one-story residence, ______________________________ plumbing may be utilized to shorten plumbing runs.

Answer: back-to-back.

Question Type: F
Random: Y
Difficulty Level: 2
Chapter Number: 15
Page Reference: 528

49. To make a house appear lower and longer, the designer can accent the _______________ lines of the elevation.

Answer: Horizontal.

Question Type: F
Random: Y
Difficulty Level: 2
Chapter Number: 15
Page Reference: 542

50. Describe a 1 1/2 story house.

Answer: A 1 1/2 story house has less living space on the top level. The useable space of the top level is cut off by the roof slope. An example would be the Cape Cod style.

Question Type: E
Random: Y
Difficulty Level: 2
Chapter Number: 15
Page Reference: 520

MULTIPLE CHOICE: Circle the letter indicating the best choice for each question.

51. A common kitchen shape is:
   a. L-shape
   b. U-shape

c. Two-wall
d. corridor
e. All of the above

Answer: e

Question Type: M
Random: Y
Difficulty Level: 2
Chapter Number: 15
Page Reference: 535

Chapters 16. BUILDING MODELS

1. What part does the model of a building play in architectural service?

Answer: Models are an important part of the architectural service because architects rely on three dimensional models for final evaluation for their designs.

Question Type: S
Random: Y
Difficulty Level: 1
Chapter Number: 16
Page Reference: 553

2. What characteristics of an architectural proposal does a model most effectively show?

Answer: The model effectively shows aspects of the building such as proportion, relationship of features, appearance of materials, and spatial concepts.

Question Type: S
Random: Y
Difficulty Level: 2
Chapter Number: 16
Page Reference: 553

3. Why is balsa wood a favorite material for many model builders?

Answer: Balsa wood is a favorite material for many model builders, since it is a soft wood that is easy to score and cut.

Question Type: S
Random: Y
Difficulty Level: 1
Chapter Number: 16
Page Reference: xx

4. How are site contours usually represented on architectural models?

Answer: Site contours are usually cut from chip board, the thickness of which represents a contour interval such as 1', 2', or 5'. Contours on large scale models may be constructed with wire screen and blocking that is covered with paper-mache.

Question Type: S

Random: Y
Difficulty Level: 2
Chapter Number: 16
Page Reference: 553

5. Why is the scale so important when selecting model building materials?

Answer: It is important to select model-building materials of the proper scale so that the model will appear realistic and be an accurate representation of the building.

Question Type: S
Random: Y
Difficulty Level: 1
Chapter Number: 16
Page Reference: 553

6. Why must the model builder use particular care as to the direction in which wood grain is placed on exterior textures?

Answer: Care must be used as to the direction in which wood grain is placed on exterior textures since the wood grain has a different appearance when it is placed vertically versus horizontally.

Question Type: S
Random: Y
Difficulty Level: 1
Chapter Number: 16
Page Reference: 564

7. Why is soft pine recommended for structural lumber rather than balsa wood?

Answer: Soft pine is recommended for structural lumber rather than balsa wood because the soft pine has more stability.

Question Type: S
Random: Y
Difficulty Level: 2
Chapter Number: 16
Page Reference: 555

8. What is the disadvantage of using glossy enamels on model surfaces?

Answer: Glossy enamel paints have the disadvantage of being reflective and emphasizing imperfections in materials and craftsmanship.

Question Type: S
Random: Y
Difficulty Level: 2
Chapter Number: 16
Page Reference: 558

9. Discuss various materials convenient for representing roofing textures.

Answer: The following materials are convenient for representing roof structures: black construction paper represents roofing felt; sand paper represents composition shingles; small wood shingles cut from an actual cedar or redwood shingle represents the wood shakes or shingles.

Question Type: E
Random: Y
Difficulty Level: 2
Chapter Number: 16
Page Reference: 564

10. Why can landscaping problems be more easily solved on a model than on a landscaping plan?

Answer: Landscape problems can be more easily solved on a model than a landscaping plan because the model has three dimensional scale and contours which only appear flat on the landscape plan.

Question Type: S
Random: Y
Difficulty Level: 2
Chapter Number: 16
Page Reference: 565

11. What is the difference between an architectural model and a structural model?

Answer: An architectural model is intended primarily to show the exterior appearance of the design only while a structural model is intended to make a visual analysis of the structural system of the building.

Question Type: S
Random: Y

Difficulty Level: 2
Chapter Number: 16
Page Reference: 553-5

12. What are some of the materials from which an architectural model may be constructed?

Answer: Some of the materials that could be used in constructing an architectural model are (1) illustration board, (2) thin balsa wood panels or (3) 1/8" thick hardboard.

Question Type: S
Random: Y
Difficulty Level: 1
Chapter Number: 16
Page Reference: 553

13. When building a structural model framing members of ____________, ____________, or ____________ are cut to scale and assembled to represent the building's actual framing system.

Answer: Balsa wood, fir or soft pine.

Question Type: F
Random: Y
Difficulty Level: 1
Chapter Number: 16
Page Reference: 555

14. What are the qualities necessary to be a successful model builder?
   a. ____________________________
   b. ____________________________
   c. ____________________________

Answer: a: Patience b: Care in attention to detail c: The ability to track down simple, everyday items that will represent building materials at a small scale.

Question Type: F
Random: Y
Difficulty Level: 2
Chapter Number: 16
Page Reference: 557

15. Name the basic tools needed for model building.

Answer: The basic tools needed for model building are:

1. model-builder's knife
2. small framing square
3. architect's scale
4. metal straightedge
5. tweezers
6. small hammer
7. handsaw
8. several small paint brushes,
9. shears
10. sandpaper and masking tape
11. a sturdy table
12. hot glue gun.

Question Type: S
Random: Y
Difficulty Level: 2
Chapter Number: 16
Page Reference: 557

16. How can brickwork be indicated on a model?

Answer: Brickwork can be indicated on a model by painting the walls with a subtle variation of the basic brick color and occasionally drawing mortar joints with white line work.

Question Type: S
Random: Y
Difficulty Level: 2
Chapter Number: 16
Page Reference: 558

17. Plaster stucco and E.I.F.S. systems can be represented by mixing sand and paint and applying this mixture as a final coat. True or False?

Answer: True

Question Type: T
Random: Y
Difficulty Level: 2
Chapter Number: 16
Page Reference: 558

18. Why should areas to receive stain be stained before gluing in place?

Answer: Areas to receive stain should be stained prior to gluing in place because the stain does not penetrate the glue and leaves bare areas where there is glue.

Question Type: S
Random: Y
Difficulty Level: 2
Chapter Number: 16
Page Reference: 564

19. What type of paints are handy for mixing paints for model building?
    a. _______________
    b. _______________
    c. _______________

Answer: a. Tempera b. acrylic c. casein, colors are handy for mixing colors for model building.

Question Type: F
Random: Y
Difficulty Level: 2
Chapter Number: 16
Page Reference: 558

20. Enamels and glossy paints will make the model look better. True or False?

Answer: False

Question Type: T
Random: Y
Difficulty Level: 2
Chapter Number: 16
Page Reference: 558

21. What is "flock"?

Answer: Flock is a tinted fiber that adheres to wet paint and may be used for ground cover on a model.

Question Type: S
Random: Y
Difficulty Level: 2
Chapter Number: 16
Page Reference: 559

22. Contours are represented by drawing lines on the model base. True or False?

Answer: False

Question Type: T
Random: Y
Difficulty Level: 1
Chapter Number: 16
Page Reference: 553

23. How can water surfaces be shown on models?

Answer: Water surfaces look realistic if represented with ripple window glass embedded in the surfacing material. Paint the area under the glass a blue-green color.

Question Type: S
Random: Y
Difficulty Level: 2
Chapter Number: 16
Page Reference: 559

24. Describe "chipboard".

Answer: Chipboard is a gray, economical paper board about 1/16" to 3/32" thick and is ideal for the making of quick, effective models. The board has good stability, takes tempera, acrylic or latex paint well and does not appear to be affected by humidity changes.

Question Type: S
Random: Y
Difficulty Level: 2
Chapter Number: 16
Page Reference: 560

25. Name three (3) materials that can be used for "massing" models.
1. ____________________
2. ____________________
3. ____________________

Answer: 1. Styrofoam blocks 2. Modeling clay 3. Soap bars

Question Type: F
Random: Y
Difficulty Level: 2
Chapter Number: 16
Page Reference: 560-2

26. What material is normally used for the base of the building?

Answer: The material that is normally used for the base of the model is 1/4" thick plywood with 1" x 2" vertical perimeter framework together with diagonal bracing on the bottom of the base.

Question Type: S
Random: Y
Difficulty Level: 2
Chapter Number: 16
Page Reference: 561

27. The dimensions for the model are taken from the ________________.

Answer: Final working drawings of the project.

Question Type: F
Random: Y
Difficulty Level: 3
Chapter Number: 16
Page Reference: 553

28. There is not any sequence in model building. True or False?

Answer: False

Question Type: T
Random: Y
Difficulty Level: 2
Chapter Number: 16
Page Reference: 560

29. Only those who specialize in model building can build models. True or False?

Answer: False

Question Type: T
Random: Y
Difficulty Level: 1
Chapter Number: 16
Page Reference: 553

30. Large areas of a model may be painted using ______________ paint.

Answer: Latex-base

Question Type: F
Random: Y
Difficulty Level: 2
Chapter Number: 16
Page Reference: 558

31. Structural models are usually built to ______________________ or ______________________ scale.

Answer: 1/2" = 1'-0" or 3/4" = 1'-0"

Question Type: F
Random: Y
Difficulty Level: 2
Chapter Number: 16
Page Reference: 555

32. The two major types of models are ______________ and ______________.

Answer: Architectural and Structural

Question Type: F
Random: Y
Difficulty Level: 1
Chapter Number: 16
Page Reference: 553

33. Architectural models are usually built to ______________________ or ______________________ scale.

Answer: 1/4" = 1'-0" or 1/8" = 1'-0"

Question Type: F
Random: Y
Difficulty Level: 2
Chapter Number: 16
Page Reference: 555

34. Discuss how to represent stonework on a model.

Answer: Stonework may be represented by painting subtle variations of basic stone colors with light-colored mortar. An alternative would be to use actual small stones.

Question Type: E
Random: Y

Difficulty Level: 2
Chapter Number: 16
Page Reference: 558

35. Chimneys may be represented with small blocks of balsa. True or False?

Answer: True

Question Type: T
Random: Y
Difficulty Level: 1
Chapter Number: 16
Page Reference: 558

36. Steep grades on a model may be made with hardware cloth covered with ________________________ or ________________________.

Answer: Plaster or Papier mache.

Question Type: F
Random: Y
Difficulty Level: 2
Chapter Number: 16
Page Reference: 559

37. Water surfaces may be represented with "ripple" window glass. True or False?

Answer: True

Question Type: T
Random: Y
Difficulty Level: 2
Chapter Number: 16
Page Reference: 560

38. Structural models should be assembled using small nails or brads. True or False?

Answer: False

Question Type: 1
Random: Y
Difficulty Level: 2

Chapter Number: 16
Page Reference: 555

39. Aluminum paint should be used to represent metal windows. True or False?

Answer: False

Question Type: T
Random: Y
Difficulty Level: 1
Chapter Number: 16
Page Reference: 564

40. Brick or stone may be constructed by scoring balsa with a(n) _____ to represent the mortar joints.

Answer: Stylus

Question Type: F
Random: Y
Difficulty Level: 2
Chapter Number: 16
Page Reference: 562

Chapters 17. WRITING SPECIFICATIONS

1. What qualifications are required of a construction specification writer?

Answer: A specification writer should have experience with construction materials, technical writing, contracts, and working drawings to qualify as a construction specification writer.

Question Type: S
Random: Y
Difficulty Level: 2
Chapter Number: 17
Page Reference: 568

2. Drawings depict the quantity of certain materials. What do specifications depict?

Answer: Specifications depict the quality of certain building materials.

Question Type: S
Random: Y
Difficulty Level: 1
Chapter Number: 17
Page Reference: 567

3. List five sources for specification reference material.

Answer: The following is a list of five sources for specification reference material: federal specifications, manufacturers' catalogs, national standards organizations, city and national codes, testing societies, master specifications, book on specs, magazines, trade associations, and files of previously written specifications.

Question Type: S
Random: Y
Difficulty Level: 2
Chapter Number: 17
Page Reference: 569

4. Why are construction material manufacturers' associations interested in furnishing materials to specification writers?

Answer: Construction material manufactures' associations are interested in furnishing information to specification writers because they hope the specifications writer will specify building products that the material manufacturers' association represents.

Question Type: S
Random: Y
Difficulty Level: 2
Chapter Number: 17
Page Reference: 569

5. Who requires a copy of the specifications? How are these copies usually made?

Answer: The architect, bidders, general contractor, sub-contractors, owner, and lending institutions require copies of specifications. Most copies of specifications are reproduced on high speed copiers.

Question Type: S
Random: Y
Difficulty Level: 2
Chapter Number: 17
Page Reference: 576

6. How is the information in the general conditions related to the other technical sections?

Answer: The information in the general conditions applies to the other technical sections?

Question Type: S
Random: Y
Difficulty Level:
Chapter Number: 17
Page Reference: 571

7. How do dimensions given in the specifications differ from the dimensions required on the working drawings?

Answer: Dimensions in specifications usually define the size of an individual unit such as a standard brick 2 1/4" x 3 3/4" x 7 5/8". Dimensions on drawings define the size of the building such as the length, width, and height.

Question Type: S
Random: Y

Difficulty Level: 2
Chapter Number: 17
Page Reference: 574

8. Why must requirements in specifications be made fair and just to all parties of the contract?

Answer: The contractor and the owner are parties to the contract for construction. Each is to perform strictly in accordance with the provisions of the contract of which the specifications are a part; therefore, the requirements in the specifications should not require either party to do things that are not fair.

Question Type: S
Random: Y
Difficulty Level: 2
Chapter Number: 17
Page Reference: 571

9. There are four (4) recommended parts in every technical section. True or False?

Answer: False
Question Type: T
Random: Y
Difficulty Level: 1
Chapter Number: 17
Page Reference: 570

10. What are the major advantages in using the five-digit numbering system in spec writing?

Answer: Discuss the major advantage of using the five-digit numbering system is that it unifies the 16 division, three-part format. The first two digits identify the division, the remaining three digits represent a sequence of sections. The specifier may create numbers for narrowscope sections using the last three digits.

Question Type: E
Random: Y
Difficulty Level: 3
Chapter Number: 17
Page Reference: 570

11. Name several architectural or construction books that follow the

five-digit numbering format.

Answer: Architectural Graphic standards and Sweets Architectural Catalog Files are both organized on the 16-division CSI format.

Question Type: S
Random: Y
Difficulty Level: 2
Chapter Number: 17
Page Reference: 570

12. What organization originated the 16-division uniform construction specification system?

Answer: The Construction specification Institute (CSI) originated the 16-division uniform system.

Question Type: S
Random: Y
Difficulty Level: 1
Chapter Number: 17
Page Reference: 570

13. Describe the difference between broadscope, mediumscope, and narrowscope?

Answer: Broadscope sections cover a series of materials which are closely related such as Acoustical ceilings." Mediumscope sections may address "lay-in acoustical ceilings" consisting of a suspension system and lay-in tiles. A narrowacope section may include veiling suspension systems."

Question Type: E
Random: Y
Difficulty Level: 3
Chapter Number: 17
Page Reference: 570

14. What is a master specification?

Answer: A master specification is a previously written specification or computerized system which has a bank of specification material stored in memory. The specifier can edit the written or computerized specification and tailor it to the need of a new construction project.

Question Type: S

Random: Y
Difficulty Level: 2
Chapter Number: 17
Page Reference: 569

15. Who generally writes specifications in a small architectural firm?

Answer: The principle who is the registered architect and owner usually writes specifications for the small firm.

Question Type: S
Random: Y
Difficulty Level: 2
Chapter Number: 17
Page Reference: 568

16. What is an allowance and why would it be used?

Answer: An allowance is an amount of money or quantity of materials for which the quantity can be determined from the working drawing but a decision as to the actual material to be used has not been selected, such as, "500 yards of carpet at $30.00 per yard is the allowance". The owner then later "selects a carpet which costs $30.00 per yard".

Question Type: E
Random: Y
Difficulty Level: 2
Chapter Number: 17
Page Reference: 575

17. What is an alternate and why would it be used?

Answer: An alternate is often included in the architects bidding documents in order to get a price for an alternative system or material from the contractor. An alternate may be "additive" or "deductive" from the contract sum, such as, "the contractor shall state the amount to be deducted from the base bid, if vinyl wall coverings are deleted and paint is substituted".

Question Type: E
Random: Y
Difficulty Level: 2
Chapter Number: 17
Page Reference: 575

18. What is a proprietary specification?

Answer: A proprietary specification is a specification which names the manufacturer or products acceptable to the architect.

Question Type: S
Random: Y
Difficulty Level: 2
Chapter Number: 17
Page Reference: 573

19. Explain how computers can aid the specification writer?

Answer: Computers aid the specification writer in achieving greater speed in specification preparation. Specification data can be taken from a CD-ROM disk and reader and stored on a personal computer hard disk. The specifier can generate an entire specification on the computer.

Question Type: S
Random: Y
Difficulty Level: 1
Chapter Number: 17
Page Reference: 576

20. The working drawings tell the contractor "____________________".
The specifications tell the contractor "________________________".

Answer: "How much" (quantities)
"How good" (quality)

Question Type: F
Random: Y
Difficulty Level: 2
Chapter Number: 17
Page Reference: xx

21. What documents compose the Contract Documents?

1. ______________________
2. ______________________
3. ______________________
4. ______________________
5. ______________________
6. ______________________

Answer:
1. Working drawings
2. Specifications
3. Contract Conditions (General and Supplemental)
4. Addenda (changes made before contract signing)

5. Change orders (changes made after contract signing)
6. Agreement (or the contract between the Owner and the Contractor)

Question Type: F
Random: Y
Difficulty Level: 3
Chapter Number: 17
Page Reference: 567

22. What is the purpose of the "Contract Documents"?

Answer: The purpose of the Contract Documents is to provide communication that will allow the Project to be constructed by the General Contractor and the sub-contractors in a fashion and quality intended by the Architect and perceived by the Owner, and as indicated in the Contract Documents.

Question Type: E
Random: Y
Difficulty Level: 3
Chapter Number: 17
Page Reference: xx

23. What documents compose the "Bidding Documents"?
   a. Invitation to bid
   b. Instructions to bidders
   c. Bid Form (proposal)
   d. Information available to bidders
   e. All of the above.

Answer: e

Question Type: M
Random: Y
Difficulty Level: 2
Chapter Number: 17
Page Reference: 567

24. Bidding documents are Contract Documents. True or False?

Answer: False

Question Type: T
Random: Y
Difficulty Level: 1
Chapter Number: 17
Page Reference: 567

25. In the case of dispute, the specifications have precedence over the drawings. True or False?

Answer: False

Question Type: T
Random: Y
Difficulty Level: 2
Chapter Number: 17
Page Reference: 567

26. What are the two (2) basic processes of determining the actual cost of construction?

Answer: The two basic methods of determining the actual cost of construction is (1) the open bid process and (2) the negotiated bid process.

Question Type: S
Random: Y
Difficulty Level: 2
Chapter Number: 17
Page Reference: 568

27. What type of delivery process would have a "guaranteed maximum cost ceiling" placed upon it?

Answer: A negotiated contract, a fast track delivery process and a design-build delivery process could have a guaranteed maximum cost ceiling.

Question Type: S
Random: Y
Difficulty Level: 2
Chapter Number: 17
Page Reference: 568

28. What is the difference between the specifications and the Project Manual?

Answer: The specifications are technical in nature while the Project Manual includes the bidding requirements, the conditions of the contract and the general construction requirements as well as the specifications.

Question Type: S
Random: Y

Difficulty Level: 2
Chapter Number: 17
Page Reference: 568

29. There are ____________________ technical divisions included in the specifications.

Answer: Fifteen (15)

Question Type: F
Random: Y
Difficulty Level: 3
Chapter Number: 17
Page Reference: 571

30. What are the four (4) types of specifications?
1. ______________________
2. ______________________
3. ______________________
4. ______________________

Answer: 1. Descriptive 2. Reference 3. Proprietary 4. Performance

Question Type: F
Random: Y
Difficulty Level: 2
Chapter Number: 17
Page Reference: 573-4

31. Describe a reference specification.

Answer: A reference specification is a specification that refers to an established standard for either a material, a test method or an installation procedure.

Question Type: E
Random: Y
Difficulty Level: 2
Chapter Number: 17
Page Reference: 574

32. In a performance specification the criteria for materials is established on the basis of physical properties of the end product. True or False?

Answer: True

Question Type: T
Random: Y
Difficulty Level: 2
Chapter Number: 17
Page Reference: 574

33. The American Institute of Architects (A.I.A.) has standardized the specification format. True or False?

Answer: False

Question Type: T
Random: Y
Difficulty Level: 2
Chapter Number: 17
Page Reference: 572

34. Describe the parts of the current specification format.

Answer: The current specification format consists of a universal 5 number system including Divisions consisting of 16 basic instructions/materials designations, Sections consisting of broadscope, mediumscope or narrowscope information of a specific material or materials and section Parts describing the specific item, its requirements, material and installation.

Question Type: E
Random: Y
Difficulty Level: 3
Chapter Number: 17
Page Reference: 572

35. An allowance is another name for unit price. True or False?

Answer: False

Question Type: T
Random: Y
Difficulty Level: 2
Chapter Number: 17
Page Reference: 575

36. How does an allowance differ from a unit price?

Answer: An allowance is price based upon the specifications for an item that will be selected at a later date. This price usually includes the cost of the material, its shipment, installation and related taxes, etc. plus the contractor's overhead and profit. A unit price is the cost to perform a certain task which a contractor proposes for each unit of that task. This price includes labor and material but does not usually include overhead and profit.

Question Type: S
Random: Y
Difficulty Level: 2
Chapter Number: 17
Page Reference: 575

37. What is a "Change Order"?

Answer: A change order is a written order for a change to the Contract signed by the Owner, Contractor and Architect which has been issued after the execution of the Agreement for construction. The change order authorizes a change in the work, an adjustment to the contract sum, an adjustment to the contract completion date or any combination of these items.

Question Type: S
Random: Y
Difficulty Level: 3
Chapter Number: 17
Page Reference: 567

38. What is the difference between an "Addendum" and a "Modification"?

Answer: An addendum is a change or correction to the drawings or specifications issued prior to the receipt of bids. A modification is a change to the construction agreement issued after the agreement has been executed. It may take the form of a field directive by the architect, a modification by the owner or architect or a corrective measure required by a field condition or error in the contract documents.

Question Type: S
Random: Y
Difficulty Level: 2
Chapter Number: 17
Page Reference: 567

39. Name the three (3) parts of a typical specification section.

a. Part I: ____________________
b. Part II: ____________________
c. Part III: ____________________

Answer: a. Part I: General, b. Part II: Products, c. Part III: Execution.

Question Type: F
Random: Y
Difficulty Level: 2
Chapter Number: 17
Page Reference: 570

40. What does Part I of a typical technical specification section include?

Answer: Part I of a typical technical specification section includes the scope of work, related items, items not included in the work, quality assurance, shop drawings required, delivery and storage and warranties.

Question Type: S
Random: Y
Difficulty Level: 2
Chapter Number: 17
Page Reference: 570

41. What does Part II of a typical technical specification section include?

Answer: Part II of a typical technical specification section includes the material description, acceptable manufacturers, mixes and fabrication of the product.

Question Type: S
Random: Y
Difficulty Level: 2
Chapter Number: 17
Page Reference: 570

42. What does Part III of a typical technical specification section include?

Answer: Part III of a typical technical specification section includes the inspection of the surfaces to receive the product and the preparation, installation and protection of the product.

Question Type: S
Random: Y
Difficulty Level: 2
Chapter Number: 17
Page Reference: 570

43. Simplify, by rewriting, the following sentence: "The general contractor is responsible for any and all decisions regarding pedestrian, supplier, personnel and automotive safety on the "project site".

Answer: The general contractor is responsible for project site safety.

Question Type: S
Random: Y
Difficulty Level: 3
Chapter Number: 17
Page Reference: xx

44. A variation of the negotiated type contract is the ______________ Type of delivery process.

Answer: Fast-track.

Question Type: F
Random: Y
Difficulty Level: 2
Chapter Number: 17
Page Reference: 568

45. In the fast-track project, construction may actually start before completion of working drawings and specs. True or False?

Answer: True

Question Type: T
Random: Y
Difficulty Level: 2
Chapter Number: 17
Page Reference: 568

46. Discuss the design/build type of project.

Answer: The design/build type of project is where the contractor is responsible for the design drawings, and construction of the project.

Question Type: E
Random: Y
Difficulty Level: 2
Chapter Number: 17
Page Reference: 568

47. A small architectural firm may hire a specification consultant to prepare their specs. True or False?

Answer: True
Question Type: T
Random: Y
Difficulty Level: 1
Chapter Number: 17
Page Reference: 569

48. An attorney should be used to write specifications. True or False?

Answer: False

Question Type: T
Random: Y
Difficulty Level: 1
Chapter Number: 17
Page Reference: 568

49. The phrase "a workmanlike job" should be included in specifications to insure quality for the project. True or False?

Answer: False

Question Type: T
Random: Y
Difficulty Level: 2
Chapter Number: 17
Page Reference: 573

50. Prices solicited from contractors per unit are referred to as, ________________________ prices.

Answer: Unit prices.

Question Type: F
Random: Y
Difficulty Level: 1
Chapter Number: 17
Page Reference: 575

51. The phrase "or equal" should be avoided when writing specifications. True or False?

Answer: True.

Question Type: T
Random: Y
Difficulty Level: 1
Chapter Number: 17
Page Reference: 573

MULTIPLE CHOICE: Circle the letter indicating the best choice for each question.

52. Which part of a technical section would include information regarding submittals?
    a. Part I
    b. Part II
    c. Part III
    d. None of the above

Answer: a

Question Type: M
Random: Y
Difficulty Level: 2
Chapter Number: 17
Page Reference: 570

53. Which part of a technical section would contain information regarding field quality control?
    a. Part I
    b. Part II
    c. Part III
    d. None of the above

Answer: c

Question Type: M
Random: Y
Difficulty Level: 2
Chapter Number: 17
Page Reference: 570

54. Which part of a technical section would contain information regarding fabrication?
    a. Part I
    b. Part II
    c. Part III

d. None of the above

Answer: b

Question Type: M
Random: Y
Difficulty Level: 2
Chapter Number: 17
Page Reference: 570

55. In what division of the Project Manual does the Bid Bond appear?
   a. Division One
   b. Division Three
   c. Division Five
   d. Division Sixteen
   e. None of the above

Answer: e

Question Type: M
Random: Y
Difficulty Level: 3
Chapter Number: 17
Page Reference: 572

Chapters 18. WORKING DRAWINGS OF A SMALL HOME

1. When starting a floor plan why is it necessary to begin with general shapes and light lines?

Answer: When starting a floor plan, one begins with general shapes and light lines to establish the overall form of the house. As the drawing progresses, adjustments may be necessary in the location of walls, windows and doors. Light lines may be easily erased.

Question Type: S
Random: Y
Difficulty Level: 1
Chapter Number: 18
Page Reference: 580

2. What is meant by a wet-wall?

Answer: A wet-wall is the wall that contains water supply pipes and sanitary drainage pipes.

Question Type: S
Random: Y
Difficulty Level: 2
Chapter Number: 18
Page Reference: 580

3. On wood frame construction drawings, why is it convenient for workers to have dimensions shown from the outside of exterior wall studs to the center lines of partitions?

Answer: On wood frame construction drawings, it is convenient for workers to have dimensions shown from the outside of exterior wall studs to the centerlines of partitions. This method is conventional, and it provides workers with the necessary dimensions for laying out the sole plates over the subfloors.

Question Type: S
Random: Y
Difficulty Level: 2
Chapter Number: 18
Page Reference: 588

4. Why are lengthy notes about construction commonly put in the specifications rather than on the working drawings?

Answer: Lengthy notes about construction may be more expediently contained in specifications which are typed rather than included in working drawings where lettering is time and space consuming.

Question Type: S
Random: Y
Difficulty Level: 2
Chapter Number: 18
Page Reference: 589

5. Why is a floor plan more readable when symbol hatching has been used throughout the walls?

Answer: A floor plan is more readable when symbol hatching has been used throughout the walls, because this gives the drawing an even tone and makes the drawing more readable by clearly defining rooms.

Question Type: S
Random: Y
Difficulty Level: 2
Chapter Number: 18
Page Reference: 589

6. Does a joist note, giving size, spacing, and direction and shown on a floor plan, pertain to the framing above or below?

Answer: The joist note giving size, spacing and direction and shown on a floor plan pertains to the framing above.

Question Type: S
Random: Y
Difficulty Level: 1
Chapter Number: 18
Page Reference: 589

7. Why are piers within a crawl space generally masonry? How is this indicated?

Answer: Piers within a crawl space are generally constructed of masonry because they are more durable than medal or wood. They are shown on the foundation plan and hatched with the proper symbol which designates the material required for the construction of the pier.

Question Type: S
Random: Y
Difficulty Level: 1

Chapter Number: 18
Page Reference: 597

8. What does the term unexcavated mean when shown on part of a foundation plan?

Answer: Unexcavated when noted on a foundation plan means the earth is not to be removed.

Question Type: S
Random: Y
Difficulty Level: 1
Chapter Number: 18
Page Reference: 597

9. Of what value is the finish floor line indication on an elevation view?

Answer: The finish floor line indication on an elevation view in of value as an important reference line so that the contractor can locate the floor at the proper position above grade.

Question Type: S
Random: Y
Difficulty Level: 2
Chapter Number: 18
Page Reference: 598

10. What dimensions must necessarily be put on an elevation view?

Answer: Dimensions on elevations need be only those that cannot be shown on details and plans. Dimension the floor and ceiling heights, chimney heights, extended wall heights, window heights, and other features requiring a vertical dimension.

Question Type: S
Random: Y
Difficulty Level: 2
Chapter Number: 18
Page Reference: 600

11. What is the minimum width for a residential driveway for one-way traffic?

Answer: The minimum width for a residential driveway for one-way traffic is 10'-0".

Question Type: S

Random: Y
Difficulty Level: 1
Chapter Number: 18
Page Reference: 602

12. At what scale are bathroom elevations usually drawn?

Answer: Bathroom elevations are usually drawn to the scale of 1/2" = 1'-0".

Question Type: S
Random: Y
Difficulty Level: 1
Chapter Number: 18
Page Reference: 601

13. What is the standard height from the floor to the top of doors in residences?

Answer: The standard height from the floor to the top of doors in a residence is 6'-8".

Question Type: S
Random: Y
Difficulty Level: 2
Chapter Number: 18
Page Reference: 598

14. Why is the floor plan developed before the foundation plan in working drawings?

Answer: The floor plan is developed before the foundation plan in working drawings because the foundation plan requires information about the total concept of the structure before it can be drawn.

Question Type: S
Random: Y
Difficulty Level: 1
Chapter Number: 18
Page Reference: 596

15. Place the following drawings in the order they would appear in a set of residential working drawings: exterior elevations, foundation plan, bathroom and kitchen details, site plan, typical wall sections, and floor plan.

Answer: The drawing in a set of residential working drawing would appear in the following order: site plan, foundation plan, floor plan, exterior elevations, typical wall sections, and bathroom and kitchen details.

Question Type: S
Random: Y
Difficulty Level: 2
Chapter Number: 18
Page Reference: 582-7

16. Why are doors numbered on a working drawing floor plan?

Answer: Doors are numbered on working drawings to cross reference the door to the corresponding number on the door schedule which indicates the type and size of the door.

Question Type: S
Random: Y
Difficulty Level: 2
Chapter Number: 18
Page Reference: 583

17. Discuss how are section cuts referenced on floor plans.

Answer: Section cuts are referenced on the floor plan by a cutting plane symbol that is shown by a bold line drawn through the point where the section is cut and a circle is attached to the cutting plane line. The circle is divided into halves by a line. A detail number or letter is placed in the upper half of the circle and a page reference number is placed in the lower half of the circle.

Question Type: E
Random: Y
Difficulty Level: 2
Chapter Number: 18
Page Reference: 583

18. What working drawing contains the most information about the construction of the residence from the foundation to the roof?

Answer: The wall section contains detailed information about the construction of the residence from the foundation to the roof.

Question Type: S
Random: Y
Difficulty Level: 2
Chapter Number: 18

Page Reference: 586

19. How far should wood siding be placed above the exterior grade at the perimeter of the building?

Answer: Wood siding should be placed 8" above the exterior grade at the perimeter of the building.

Question Type: S
Random: Y
Difficulty Level: 2
Chapter Number: 18
Page Reference: 586

20. Name four (4) types of drawings included in a set of residential working drawings.

a. ____________________
b. ____________________
c. ____________________
d. ____________________

Answer: a. Floor plans b. Exterior elevations c. Sections and d. Details are the major ones.

Question Type: F
Random: Y
Difficulty Level: 1
Chapter Number: 18
Page Reference:

21. Layout of floor plans should start with the using a F-pencil. True or False?

Answer: False

Question Type: T
Random: Y
Difficulty Level: 1
Chapter Number: 18
Page Reference: 580

22. What are satisfactory dimensions for the following wall types to be used when laying out the floor plans?

a. Wood frame exterior walls ____________
b. Wood frame interior partitions ____________

c. Brick veneer on wood frame __________
d. Brick cavity with 2" air space ___________

Answer: a. 6" b. 5" c. 10" d. 10"

Question Type: F
Random: Y
Difficulty Level: 3
Chapter Number: 18
Page Reference: 580

23. What information must be known before the stairs may be drawn on the floor plans?

Answer: The total number of treads in the run of the stairs (this will be determined by the total floor-to-floor height) must be known before the stairs can be drawn on the floor plans.

Question Type: S
Random: Y
Difficulty Level: 3
Chapter Number: 18
Page Reference: 588

24. What are the notations that are required on the stair drawings?

Answer: The notations required on the stair drawings are an arrow indicating the direction of the stair and the number of risers in the total run of the stairs. (In addition, the notation for a railing is sometimes shown.)

Question Type: S
Random: Y
Difficulty Level: 2
Chapter Number: 18
Page Reference: 588

25. What is the usual size of a fireplace hearth?

Answer: The fireplace hearth usually extends 18" to 20" into the room from the front of the fireplace and 12" to either side of the fireplace opening. (Local codes determine these dimensions).

Question Type: S
Random: Y
Difficulty Level: 2

Chapter Number: 18
Page Reference: 588

26. The hearth is usually supported by the sub-floor. True or False?

Answer: False

Question Type: T
Random: Y
Difficulty Level: 2
Chapter Number: 18
Page Reference: 588

27. Kitchen base cabinets are indicated ____________________ deep and wall cabinets are indicated ________________________ deep.

Answer: Base cabinets are indicated 24" deep and wall cabinets are indicated 12" deep.

Question Type: F
Random: Y
Difficulty Level: 1
Chapter Number: 18
Page Reference: 588

28. Why is it important to draw the bathroom and kitchen fixtures accurately to scale?

Answer: It is important to draw the kitchen and bathroom fixtures to scale to indicate correct placement and workability of the arrangement.

Question Type: S
Random: Y
Difficulty Level: 2
Chapter Number: 18
Page Reference: 588

29. Permanent built-in features do not have to be indicated on the floor plan but may be drawn as details on another sheet. True or False?

Answer: False

Question Type: T
Random: Y

Difficulty Level: 2
Chapter Number: 18
Page Reference: 588

30. How is wiring circuitry indicated on the floor plan?

Answer: Wiring circuitry is indicated on the floor plan by a broken line connecting one device to another.

Question Type: S
Random: Y
Difficulty Level: 2
Chapter Number: 18
Page Reference: 588

31. What is the purpose of a three-way lighting switch?

Answer: The purpose of a three-way lighting switch is to allow the lights of the area to be turned off and on from two remote locations.

Question Type: S
Random: Y
Difficulty Level: 2
Chapter Number: 18
Page Reference: 588

32. Duplex wall outlets are usually spaced __________________ in residential living areas.

Answer: Eight feet (8'-0")

Question Type: F
Random: Y
Difficulty Level: 2
Chapter Number: 18
Page Reference: 588

33. In traditional wood frame construction where do the exterior overall dimensions begin and end?

Answer: In traditional wood frame construction the exterior dimensions begin and end at the outside surface of the exterior stud wall at each end of the house.

Question Type: S
Random: Y

Difficulty Level: 2
Chapter Number: 18
Page Reference: 588

34. In traditional wood frame construction how are exterior windows and doors dimensioned?

Answer: Windows and doors are dimensioned to their center lines in traditional wood frame construction.

Question Type: S
Random: Y
Difficulty Level: 2
Chapter Number: 18
Page Reference: 588

35. In relation to the building where are the following exterior dimensions placed on residential working drawings?
   a. Window and door openings ______________________
   b. Major breaks in the building ______________________
   c. Overall building dimensions ______________________

Answer: a: Closest to the building b: Second from the outermost dimension c: The outermost dimension

Question Type: F
Random: Y
Difficulty Level: 3
Chapter Number: 18
Page Reference: 588

36. Dimension figures and notes should not be any larger than _______ high.

Answer: One-eight inch (1/8")

Question Type: F
Random: Y
Difficulty Level: 1
Chapter Number: 18
Page Reference: 588

37. What information does a "Finish Schedule" contain?

Answer: A finish schedule contains the materials, finishes and other pertinent information regarding each space in the project. The spaces are cross-referenced to the floor plan by a system of names and, on most working drawings, numbers.

Question Type: S
Random: Y
Difficulty Level: 2
Chapter Number: 18
Page Reference: 592

38. What information is indicated on the "Door Schedule"?

Answer: When these schedules are used the door schedule will contain information concerning the type, material, size, thickness, number and swing of the doors, the type, material and detail location of the door frame, fire rating, undercut required and threshold. The door or opening schedule is cross-referenced to the floor plan by the door or opening numbers.

Question Type: S
Random: Y
Difficulty Level: 2
Chapter Number: 18
Page Reference: 594

39. What is the purpose of the hatching of materials ?

Answer: The purpose of hatching the materials is to indicate the various materials used in the construction of the Project and their location and relationship to each other.

Question Type: S
Random: Y
Difficulty Level: 2
Chapter Number: 18
Page Reference: 589

40. Why are residential bedroom doors a minimum of 2-6" wide?

Answer: The 2'-6" dimension is the smallest width that will easily allow furniture to be moved into and out of a room. (Some odes also require)

Question Type: S
Random: Y
Difficulty Level: 2
Chapter Number: 18
Page Reference: 584

41. How are the footings indicated on the foundation plan?

Answer: The footings are indicated on the foundation plan by dashed lines since they are below the grade and, therefore, are hidden from view.

Question Type: S
Random: Y
Difficulty Level: 2
Chapter Number: 18
Page Reference: 597

42. Joist information, when shown on the first floor plan, gives the size, spacing and direction of the first floor joists. True or False?

Answer: False

Question Type: T
Random: Y
Difficulty Level: 2
Chapter Number: 18
Page Reference: 589

43. What information is generally indicated on the exterior elevations?

Answer: Vertical dimensions are generally indicated on the exterior elevations.

Question Type: S
Random: Y
Difficulty Level: 2
Chapter Number: 18
Page Reference: 600

44. At what scale are kitchen elevations drawn?

Answer: Kitchen elevations are drawn at 1/2' = 1'-0" scale.

Question Type: S
Random: Y
Difficulty Level: 2
Chapter Number: 18
Page Reference: 601

45. How can columns and load bearing walls be eliminated from

residential basements and still retain the structural integrity of the floor above?

Answer: Columns and load bearing walls may be eliminated from residential basements by incorporating either clear span steel beams of a size sufficient to carry the required loads or by using pre-fabricated wood joists or "truss-joists" to span from one perimeter foundation wall to the opposite foundation wall.

Question Type: S
Random: Y
Difficulty Level: 2
Chapter Number: 18
Page Reference: 586

46. The basement or foundation plan should be developed before the first floor plan. True or False?

Answer: False

Question Type: F
Random: Y
Difficulty Level: 1
Chapter Number: 18
Page Reference: 596

47. In residential design and construction the standard height from the floor to the top of the window is ____________.

Answer: Six feet, eight inches (6'-8")

Question Type: F
Random: Y
Difficulty Level: 3
Chapter Number: 18
Page Reference: 598

48. Give the sheet designations and numbers for the following residential working drawings:

______ a. Exterior elevations
______ b. Foundation plan
______ c. Bathroom and kitchen details and elevations
______ d. Site plan
______ e. Typical wall sections
______ f. Floor plan (single story residence)

Answer: a: A4 b: A2 c: A6 d: A1 e: A5 f: A3

Question Type: F
Random: Y
Difficulty Level: 3
Chapter Number: 18
Page Reference: 582-7

49. What drawing contains the most information about the construction of the residence from the foundation to the roof?

Answer: The floor plan contains the most information about the construction of the residence from the foundation to the roof. True or False?

Question Type: T
Random: Y
Difficulty Level: 2
Chapter Number: 18
Page Reference: 586

50. Wood siding should be placed at the exterior grade at the perimeter of the building. True or False?

Answer: False

Question Type: T
Random: Y
Difficulty Level: 1
Chapter Number: 18
Page Reference: 586

51. Sketch the symbol for a 3-way switch.

Answer: (See page 581)

Question Type: S
Random: Y
Difficulty Level: 1
Chapter Number: 18
Page Reference: 581

52. Sketch the symbol for a 220 range outlet.

Answer: (See page 581)

Question Type: S
Random: Y
Difficulty Level: 2
Chapter Number: 18
Page Reference: 581

53. Sketch the symbol for a dryer outlet.

Answer: (See page 581)

Question Type: S
Random: Y
Difficulty Level: 2
Chapter Number: 18
Page Reference: 581

54. Sketch the symbol for a convenience outlet.

Answer: (See page 581)

Question Type: S
Random: Y
Difficulty Level: 2
Chapter Number: 18
Page Reference: 581

55. Sketch the symbol for a telephone jack.

Answer: (See page 581)

Question Type: S
Random: Y
Difficulty Level: 2
Chapter Number: 18
Page Reference: 581

56. Sketch the symbol for frame lumber (in section).

Answer: (See page 581)

Question Type: S
Random: Y
Difficulty Level: 2
Chapter Number: 18
Page Reference: 581

57. Sketch the symbol for finish lumber (in section).

Answer: (See page 581)

Question Type: S
Random: Y
Difficulty Level: 2
Chapter Number: 18
Page Reference: 581

58. Describe a transverse section.

Answer: A transverse section is a full section cut thru the narrowest dimension of a building.

Question Type: S
Random: Y
Difficulty Level: 2
Chapter Number: 18
Page Reference: 600

59. To indicate room dimensions , the architect may choose to dimension to the wall face instead of the center of walls. True or False?

Answer: True.

Question Type: T
Random: Y
Difficulty Level: 2
Chapter Number: 18
Page Reference: 588

60. Room titles should be placed ________________________ of the room on a floor plan.

Answer: near the center

Question Type: F
Random: Y
Difficulty Level: 2
Chapter Number: 18
Page Reference: 589

MULTIPLE CHOICE: Circle the letter indicating the best choice for each question.

61. What is the normal width of a residential entrance door?

a. 2'-4"
b. 2'-6"
c. 2'-8"
d. 3'-0"
e. 3'-6"

Answer: d

Question Type: M
Random: Y
Difficulty Level: 2
Chapter Number: 18
Page Reference: 584 & 594

1. How can the architecture contribute to the efficiency of the heating-cooling systems?

Answer: Some of the ways the architecture can contribute to the efficiency of the heating and cooling system are to protect the west wall from solar radiation; place windows on the south side; specify a light colored roof; design the house with a barrier to protect against the winter winds; specify weather stripping, insulated glass, and adequate insulation.

Question Type: E
Random: Y
Difficulty Level: 2
Chapter Number: 19
Page Reference: 606

2. Usually ___________ degrees F, is taken as a comfortable inside temperature in winter.

Answer: 72

Question Type: F
Random: Y
Difficulty Level: 1
Chapter Number: 19
Page Reference: 618

3. What are the advantages of the forced warm-air heating systems?

Answer: The advantage of forced air is mainly the economical initial cost and its compatibility with cool air systems. Heat distribution through out the house is almost instantaneous and the system is not restrictive in its location.

Question Type: S
Random: Y
Difficulty Level: 2
Chapter Number: 19
Page Reference: 607

4. What are the advantages of the hydronic heating systems?

Answer: The advantage of the hydroponic system is that the same boiler can be used to heat water for domestic use as well as for space heating.

Question Type: S
Random: Y
Difficulty Level: 2
Chapter Number: 19
Page Reference: 609

5. What are the advantages of the radiant heating systems?

Answer: The advantage of the system is even distribution of heat, no noise, and the pipes can be concealed in floors, walls or ceilings.

Question Type: S
Random: Y
Difficulty Level: 2
Chapter Number: 19
Page Reference: 615

6. Would a heat-pump system be efficient for heating a home in your locality?

Answer: The most practical adoption of the heat pump for home heating is in regions where the outside temperatures seldom drop below 30 degrees F.

Question Type: S
Random: Y
Difficulty Level: 2
Chapter Number: 19
Page Reference: 616

7. Why are filters placed in the return-duct side of a warm-air heating system?

Answer: The filters are placed on the return duct side of the warm air heating system to keep dust from collecting in the heat exchange unit.

Question Type: S
Random: Y
Difficulty Level: 2
Chapter Number: 19
Page Reference: 608

8. What type of warm-air furnace must be used with heat ducts embedded in a concrete slab-on-ground floor?

Answer: When ducts are put in concrete floor. or on the ground floor, a counter-flow system must be used to push the heat down into the ducts.

Question Type: S
Random: Y
Difficulty Level: 2
Chapter Number: 19
Page Reference: 610

9. How is air removed from kitchens and bathrooms?

Answer: The air lost from the house because of vent fans in the bathrooms and kitchens is now being replaced with fresh air in return ducts to equalize return air pressures and keep a continual supply of fresh air in the house.

Question Type: S
Random: Y
Difficulty Level: 1
Chapter Number: 19
Page Reference: 606

10. What three controls are necessary in making a warm-air system fully automatic?

Answer: To have a fully automatic warm air heating system one must have the following controls: room thermostat, fan thermostat, and a high limit control switch.

Question Type: S
Random: Y
Difficulty Level: 2
Chapter Number: 19
Page Reference: 608

11. Discuss why has electricity become popular as a home heating energy source.

Answer: Electricity has become a very popular heating source where electricity is economical; it is clean heat that is very easy to control and distribute.

Question Type: E

Random: Y
Difficulty Level: 2
Chapter Number: 19
Page Reference: 615

12. Why would the 2-pipe hot water system be more satisfactory than the 1-pipe system in home heating?

Answer: The two pipe hot water system would be more desirable because the system produces more even heat throughout the house.

Question Type: S
Random: Y
Difficulty Level: 2
Chapter Number: 19
Page Reference: 612

13. What general type of insulation is the most efficient with radiant heating systems?

Answer: A good type of insulation to use with radiant heating is a reflective insulation.

Question Type: S
Random: Y
Difficulty Level: 2
Chapter Number: 19
Page Reference:

14. What two methods of home heating commonly employ baseboard registers?

Answer: The two methods of home heating that commonly employ base board registers are hot water systems and electric strip heat systems.

Question Type: S
Random: Y
Difficulty Level: 2
Chapter Number: 19
Page Reference: 609-15

15. Which type of heating system is most easily combined with conventional cooling equipment?

Answer: The forced warm air systems are most suitable when combined with a cooling system.

Question Type: S
Random: Y
Difficulty Level: 1
Chapter Number: 19
Page Reference: 607

16. If a room has a calculated heat loss of 3,200 Btu per hr., what wattage resistance wire would be needed to satisfy the heat?

Answer: 3,200 Btu ÷ 3.41 Btu per watt = 938.42 watts resistance wiring is needed.

Question Type: S
Random: Y
Difficulty Level: 3
Chapter Number: 19
Page Reference: 615

17. In heat-pump installation, what is the primary source of heat supplied to the rooms?

Answer: Exterior heat is transferred through a heat pump to heat the interior.

Question Type: S
Random: Y
Difficulty Level: 2
Chapter Number: 19
Page Reference: 616

18. What mechanism reverses the cycles from heating to cooling in the heat pump?

Answer: To change a heat pump from heating to cooling you must reverse the refrigeration cycle.

Question Type: S
Random: Y
Difficulty Level: 2
Chapter Number: 19
Page Reference: 616

19. What is the purpose of an outside thermostat?

Answer: The outside thermostat keeps the radiant heating system from getting too far behind.

Question Type: S
Random: Y
Difficulty Level: 2
Chapter Number: 19
Page Reference: 615

20. What are the disadvantages of water chillers in residential cooling?

Answer: The main problem with water chillers for residential use is that their size, complexity and demand for power make them uneconomical.

Question Type: S
Random: Y
Difficulty Level: 3
Chapter Number: 19
Page Reference: 618

21. Partitions between heated rooms should be calculated at one-half the normal heat losses. True or False?

Answer: False

Question Type: T
Random: Y
Difficulty Level: 1
Chapter Number: 19
Page Reference: 619

22. Discuss how initial plumbing costs be reduced when designing a home?

Answer: The best way to lower plumbing costs is to centralize all plumbing in one wall if possible and keep the fixtures as close to each other as possible.

Question Type: E
Random: Y
Difficulty Level: 3
Chapter Number: 19
Page Reference:

23. What size hot-water heater would be sufficient for a family of three, where the home contains automatic laundry equipment?

Answer: The average family needs a 30 gallon hot water tank.

Question Type: S
Random: Y
Difficulty Level: 2
Chapter Number: 19
Page Reference: 635

24. What factors determine the correct size septic tank and the length of the tile drainage lines?

Answer: The size of the septic tank is determined by local codes according to the size of the dwelling and the number of occupants. The length of the tile drainage lines is based on the percolation rate of the soil.

Question Type: S
Random: Y
Difficulty Level: 2
Chapter Number: 19
Page Reference: 635

25. The minimum size electrical service panel that should be selected today for a home of 2,000 sq. ft. is _______________. Explain.

Answer: A 2,000 square foot house would need a 100 ampere service; if the house has more major electrical appliances a 150 ampere service; and if the house has electric heating it will need a 200 ampere service.

Question Type: E
Random: Y
Difficulty Level: 2
Chapter Number: 19
Page Reference: 639

26. What is a "feeder" circuit?

Answer: A feeder circuit connects power from the main power panel to sub-panel.

Question Type: S
Random: Y
Difficulty Level: 2
Chapter Number: 19
Page Reference: 640

27. What should be analyzed to determine the location of wall switches?

Answer: When placing wall switches, one factor to be considered would be that switches are generally placed 4' above the floor in the direction of the flow of traffic so one can turn lights off and on when leaving or entering a room.

Question Type: E
Random: Y
Difficulty Level: 2
Chapter Number: 19
Page Reference: 642

28. What types of switches are needed to operate an outlet from three locations?

Answer: Two three-way switches, and one four-way switch.

Question Type: S
Random: Y
Difficulty Level: 2
Chapter Number: 19
Page Reference: 643

29. Why must lighting be considered in the total concept of a home?

Answer: Well lighted interiors are often decidedly more interesting in the evening than during the day light. Artificial lighting must be regarded as an important part of the total design of the building.

Question Type: E
Random: Y
Difficulty Level: 2
Chapter Number: 19
Page Reference: 641

30. Registered engineers are required to design the mechanical and electrical systems in residential construction. True or False?

Answer: False

Question Type: T
Random: Y
Difficulty Level: 2
Chapter Number: 19

Page Reference:

31. Why is the selection of the most appropriate heating/cooling system important in residential design and construction?

Answer: The selection of the most appropriate heating/cooling system is important in residential design and construction because the system is closely related to other aspects of the design, i.e. the necessity of a utility room, the feasibility of a basement, selection of the floor construction, etc. and requires more consideration during early planning than other equipment installations.

Question Type: E
Random: Y
Difficulty Level: 2
Chapter Number: 19
Page Reference:

32. Name five (5) considerations in the selection of a heating/cooling system.
1. ____________________________
2. ____________________________
3. ____________________________
4. ____________________________
5. ____________________________

Answer: Some of the considerations in the selection of the heating/cooling system are: (1) Type and cost of fuel available, (2) Client's preference to type of heating/cooling, (3) Local climate, (4) Degree of comfort desired, (5) Initial cost of equipment, (6) Efficiency of equipment, (7) Amount of insulation required and (8) Is cooling a necessity in that location.

Question Type: F
Random: Y
Difficulty Level: 3
Chapter Number: 19
Page Reference: 606

33. Name the four (4) types of residential heating systems.
1. ____________________________
2. ____________________________
3. ____________________________
4. ____________________________

Answer: 1. Forced warm air system. 2. Hot-water system.

3. Electric resistance system. 4. Heat pump system

Question Type: F
Random: Y
Difficulty Level: 2
Chapter Number: 19
Page Reference: 606

34. What type of ductwork can be used with warm air furnaces?

Answer: Rectangular or circular sheet metal ducts, rectangular or circular fiberglass ducts and circular PVC coated steel ducts may be used with warm air furnaces.

Question Type: S
Random: Y
Difficulty Level: 2
Chapter Number: 19
Page Reference: 608

35. Describe the "Perimeter Arrangement" of ductwork.

Answer: The perimeter arrangement of ductwork places a continuous duct near the periphery of the building with a number of "feeder" ducts radiating out from the central plenum supplying the perimeter duct with warm air. Room registers receive the warm air from the perimeter duct only.

Question Type: E
Random: Y
Difficulty Level: 3
Chapter Number: 19
Page Reference: 609

36. What equipment comprises the modern warm-air heating system?

Answer: The furnace, ductwork, fan, registers and controls comprise the modern warm-air heating system.

Question Type: S
Random: Y
Difficulty Level: 3
Chapter Number: 19
Page Reference: 606-611

37. What is a "Humidifier" and how does it operate?

Answer: A humidifier is a device used on a warm-air furnace which adds water vapor to dry air and is controlled by a "humidistat" or humidity detector which sets the correct humidity level.

Question Type: S
Random: Y
Difficulty Level: 2
Chapter Number: 19
Page Reference: 608

38. Where should the room thermostat be located?

Answer: The room thermostat should be located on an interior wall and be placed approximately five feet (5'-0") above the floor.

Question Type: S
Random: Y
Difficulty Level: 2
Chapter Number: 19
Page Reference: 608

39. Describe a Hot-water heating system.

Answer: The hot-water heating system has a central boiler utilizing gas, oil or coal as a fuel and heats the water from a cold water supply to the required temperature, pumps the conditioned water throughout the rooms by means of various piping systems.

Question Type: E
Random: Y
Difficulty Level: 3
Chapter Number: 19
Page Reference: 609

40. What is the difference between a one-pipe and a two-pipe hot-water heating system?

Answer: The one-pipe hot-water heating system utilizes a single pipe running continuous from the boiler throughout the rooms and back to the boiler, thus being a combination supply and return piping system. The two-pipe hot-water heating system utilizes both a supply and a separate return piping system. The return system can be either direct or reverse with the reverse-return being the preferred.

Question Type: E
Random: Y

Difficulty Level: 3
Chapter Number: 19
Page Reference: 612

41. Name the possible locations of the piping and room equipment for a hot-water heating system.

Answer: The possible locations of the piping and room equipment for a hot-water heating system are (1) Embedded in either the concrete floor slab or the plaster ceiling, (2) Conventional radiators and (3) Baseboard convection units.

Question Type: S
Random: Y
Difficulty Level: 3
Chapter Number: 19
Page Reference: 612

42. What is the purpose of an expansion tank in a hot-water heating system?

Answer: The purpose of an expansion tank in a hot-water heating system is to compensate for variations of the water volume at different temperature levels and to relieve air pockets from the system.

Question Type: S
Random: Y
Difficulty Level: 2
Chapter Number: 19
Page Reference: 612

43. What are the advantages of a "radiant-panel" heating system?

a. ______________________________
b. ______________________________
c. ______________________________

Answer: a. The heat is "clean" creating little dust and reducing the need for frequent painting. b. There are not any visible registers or heating units within the rooms. c: There are no drafts or noise from blower equipment.

Question Type: F
Random: Y
Difficulty Level: 2
Chapter Number: 19
Page Reference: 615

44. Electric heating systems convert electrical energy into heat with the use of resistance wiring installed in various ways throughout the structure. True or False?

Answer: True

Question Type: T
Random: Y
Difficulty Level: 2
Chapter Number: 19
Page Reference: 615

45. 1 watt = __________ Btu.

Answer: 3.41

Question Type: F
Random: Y
Difficulty Level: 1
Chapter Number: 19
Page Reference: 615

46. What is one of the architectural advantages of electrical resistance-type heat?

Answer: One of the architectural advantages of electrical resistance-type heat is that floor space does not have to be reserved for a central heating unit nor is there a need for a chimney or flue.

Question Type: S
Random: Y
Difficulty Level: 2
Chapter Number: 19
Page Reference: 615

47. Discuss the theory of the "Heat pump".

Answer: The heat pump does not use gas, oil or coal as a primary heat source but operates on the principal that freely available outside air contains useful heat even during colder weather and that this heat can be pumped into the house to maintain desirable temperature levels. In summer, the system operates in reverse and takes the heat out of the house, working on a principal similar to that of the residential refrigerator.

Question Type: E

Random: Y
Difficulty Level: 3
Chapter Number: 19
Page Reference: 616

48. The heat pump operates efficiently at all winter temperatures. True or False?

Answer: False.
Question Type: T
Random: Y
Difficulty Level: 1
Chapter Number: 19
Page Reference: 616

49. ______________________________, ____________________________________,

and _________________________ are proven successful heat source-and-delivery media for a heat pump.

Answer: Water-to water, water-to-air and air-to-air

Question Type: F
Random: Y
Difficulty Level: 2
Chapter Number: 19
Page Reference: 616

50. What are the normal areas of a residence where ventilation problems occur?

Answer: Ventilation problems normally occur in the kitchen, bathroom, laundry, basement and attic areas of a residence.

Question Type: S
Random: Y
Difficulty Level: 2
Chapter Number: 19
Page Reference: 618

51. How are ventilation problems resolved in a residence?

Answer: Ventilation problems are resolved in a residence by the use of individual exhaust fans located in each area.

Question Type: S
Random: Y

Difficulty Level: 2
Chapter Number: 19
Page Reference: 618

52. Sixty-five degrees is considered to be a comfortable inside temperature for a residence. True or False?

Answer: False.

Question Type: T
Random: Y
Difficulty Level: 1
Chapter Number: 19
Page Reference: 618

53. How is the "Design temperature differences" calculated?

Answer: The design temperature difference is calculated by subtracting the recommended outside temperature (varies throughout the country) from the comfortable inside temperature.

Question Type: S
Random: Y
Difficulty Level: 2
Chapter Number: 19
Page Reference: 618

54. What is the "U-factor" in calculating heat loss?

Answer: The U-factor is the coefficient of transmission through a combination of materials commonly found in walls, floors and ceilings.

Question Type: S
Random: Y
Difficulty Level: 2
Chapter Number: 19
Page Reference: 619

55. What is the "k-factor" in calculating heat loss?

Answer: The k-factor is the rate of heat flow in Btuh through one square foot of homogeneous material 1" thick per degree (F) difference in temperature between its two surfaces.

Question Type: S
Random: Y

Difficulty Level: 2
Chapter Number: 19
Page Reference: 619

56. What does the "R-factor" represent in calculating heat loss and how is it obtained?

Answer: The R-factor represents the thermal resistance of a material. It is obtained from the reciprocal of U factors and is expressed in degrees (F) per Btuh per square foot.

Question Type: S
Random: Y
Difficulty Level: 2
Chapter Number: 19
Page Reference: 619

57. The larger the "R-factor", the more heat loss. True or False?

Answer: False

Question Type: T
Random: Y
Difficulty Level: 1
Chapter Number: 19
Page Reference: 619

58. The rate at which a building loses heat depends directly upon the heat-loss characteristics of its structural components and upon the temperature gradient. True or False?

Answer: True

Question Type: T
Random: Y
Difficulty Level: 1
Chapter Number: 19
Page Reference: 619

59. Are partitions between heated rooms considered when calculating heat loss?

Answer: Partitions between heated rooms are not considered when calculating heat loss.

Question Type: S
Random: Y

Difficulty Level: 1
Chapter Number: 19
Page Reference: 619

60. What is the purpose of a solar collector?

Answer: The purpose of a solar collector is to convert incident solar radiation into useful thermal energy.

Question Type: S
Random: Y
Difficulty Level: 2
Chapter Number: 19
Page Reference: 623

61. Three (3) types of solar collectors are:
1.________________________
2.________________________
3.________________________

Answer: 1. Flat-plate collector 2. Air-cooled collector 3. Liquid-cooled collector.

Question Type: F
Random: Y
Difficulty Level: 2
Chapter Number: 19
Page Reference: 624

62. What are the three (3) general areas of plumbing that the plumbing drawings cover?

Answer: The three general areas that the plumbing drawings cover are the domestic hot and cold water supply lines, the sanitary disposal piping system, and the plumbing fixtures.

Question Type: S
Random: Y
Difficulty Level: 2
Chapter Number: 19
Page Reference: 630

63. Which part of the plumbing system operates by pressure?

Answer: The parts of the plumbing system that operates by pressure are the domestic hot and cold water supply lines.

Question Type: S
Random: Y
Difficulty Level: 2
Chapter Number: 19
Page Reference: 632

64. Which part of the plumbing system operates by gravity?

Answer: The part that operates by gravity is the sanitary disposal system.

Question Type: S
Random: Y
Difficulty Level: 2
Chapter Number: 19
Page Reference: 632

65. What is the purpose of "stacking" the plumbing?

Answer: The purpose of stacking the plumbing is to reduce the initial plumbing cost of the Project.

Question Type: S
Random: Y
Difficulty Level: 2
Chapter Number: 19
Page Reference: 634

66. What fittings are required at the water service entrance to a residence?

a.____________________
b.____________________
c.____________________
d.____________________

Answer: a. Shutoff valve b. Pressure reducing valve c. Backflow preventer d. Pressure relief valve

Question Type: F
Random: Y
Difficulty Level: 2
Chapter Number: 19
Page Reference: 632

67. What is the purpose of the trap seal?

Answer: The purpose of the trap seal is to prevent sewer gases from entering the building from the sanitary disposal system.

Question Type: S
Random: Y
Difficulty Level: 3
Chapter Number: 19
Page Reference: 634

68. What is the purpose of vents in the plumbing system?

Answer: The purpose of the vents in the plumbing system is to permit atmospheric pressure on both sides of the trap seal to be maintained and to permit air to enter the system at the same time the gases are being removed.

Question Type: S
Random: Y
Difficulty Level: 3
Chapter Number: 19
Page Reference: 634

69. What are the conditions affecting the design of a private sanitary disposal system?

Answer: The conditions affecting the design of a private sanitary disposal system are (1) number of occupants of the house, (2) permeability of the soil, (3) groundwater level in the area, (4) size of the property, and (5) relative location of wells or streams that may be affected by the sewage disposal.

Question Type: S
Random: Y
Difficulty Level: 3
Chapter Number: 19
Page Reference: 635

70. The standard size for a domestic water heater is 100 gallons. True or False?

Answer: False

Question Type: T
Random: Y
Difficulty Level: 1
Chapter Number: 19
Page Reference: 635

71. What is the minimum electric service that may be installed in a residence according to the National Electric Code?

Answer: The minimum electric service that may be installed in a residence is sixty (60) ampere service.

Question Type: S
Random: Y
Difficulty Level: 2
Chapter Number: 19
Page Reference: 639

72. Describe "nonmetallic-sheathed" electric cable.

Answer: Nonmetallic-sheathed electric cable is a composite of individual insulated flexible copper wires covered with a paper and plastic that form the protective coating.

Question Type: S
Random: Y
Difficulty Level: 2
Chapter Number: 19
Page Reference: 639

73. What is the disadvantage of "metal-armored" electric cable?

Answer: Although "metal-armored" cable is covered with a spiral, flexible-metal sheathing that gives the cable more protection from physical damage, its disadvantage is that it is unsuitable for use in damp areas.

Question Type: S
Random: Y
Difficulty Level: 2
Chapter Number: 19
Page Reference: 640

74. What is the maximum distance apart that receptacle outlets should be placed in a residence?

Answer: Receptacle outlets in a residence should be spaced so that no point along a wall shall be more than six (6) feet from an outlet.

Question Type: S
Random: Y
Difficulty Level: 2
Chapter Number: 19

Page Reference: 640

75. Name four (4) pieces of equipment that require individual circuits.

Answer: Equipment that require individual circuits are (1) electric range, (2) washer-dryer, (3) fuel-fired furnace, (4) dishwasher, (5) electric water heater, (6) central air conditioning, (7) bathroom heaters and other similar equipment.

Question Type: S
Random: Y
Difficulty Level: 3
Chapter Number: 19
Page Reference: 638

76. Where are "ground fault" receptacles required?

Answer: Ground fault receptacles are required in bathrooms, kitchens and all outside areas.

Question Type: S
Random: Y
Difficulty Level: 2
Chapter Number: 19
Page Reference: 640

77. Where should wall lighting switches be located?

Answer: Wall lighting switches should be located near the latch side of door openings and be four (4) feet above the floor.

Question Type: S
Random: Y
Difficulty Level: 2
Chapter Number: 19
Page Reference: 642

78. Define AFUE (annual fuel-utilization efficiency).

Answer: AFUE (annual fuel-utilization efficiency)is the measure of a furnace or boilers's efficiency averaged over a heating season.

Question Type: S
Random: Y
Difficulty Level: 2

Chapter Number: 19
Page Reference: 607

79. _____________ is the ratio of the amount of energy delivered to the amount of energy consumed.

Answer: COP (coefficient of performance)

Question Type: F
Random: Y
Difficulty Level: 2
Chapter Number: 19
Page Reference: 607

80. _____________ is the ratio of seasonal cooling performance of heat pumps and air conditioners in BTU's, divided by the energy consumption in watt-hours, for an average U.S. climate.

Answer: SEER (seasonal energy-efficient rating).

Question Type: F
Random: Y
Difficulty Level: 2
Chapter Number: 19
Page Reference: 607

81. ___________ is the ratio of the estimated seasonal heating output in BTU's, divided by seasonal power consumption in watts; used for comparing air-source heat pump performance.

Answer: HSPF (heating-season performance factor)

Question Type: F
Random: Y
Difficulty Level: 2
Chapter Number: 19
Page Reference: 607

82. Discuss the placement of outlets in a room for forced air heating systems.

Answer: Generally, it is best to place outlets near outside walls where the greatest amount of heat loss or heat gain occurs.

Question Type: E
Random: Y
Difficulty Level: 2

Chapter Number: 19
Page Reference: 621

83. With solar heating, other means of heating in a house are eliminated. True or False?

Answer: False

Question Type: T
Random: Y
Difficulty Level: 1
Chapter Number: 19
Page Reference: 627

84. A safety devise that contains an electrical conductor that melts when the circuit is overloaded, is a(n) ________________.

Answer: Fuse

Question Type: F
Random: Y
Difficulty Level: 1
Chapter Number: 19
Page Reference: 637

85. A(n) ________________ is a point at which electricity serves lights, appliances, and other equipment.

Answer: Outlet

Question Type: F
Random: Y
Difficulty Level: 2
Chapter Number: 19
Page Reference: 637

86. A(n) ________________ is where two or more conductors carry electricity from a source to one or more outlets and then returns.

Answer: Circuit

Question Type: F
Random: Y
Difficulty Level: 2
Chapter Number: 19
Page Reference: 637

87. Discuss the advantages of circuit breakers over fuses.

Answer: Circuit breakers do not require replacement when the breaker has tripped. A fuse would have to be replaced.

Question Type: E
Random: Y
Difficulty Level: 2
Chapter Number: 19
Page Reference: 637

88. Sketch the symbol used to show a ceiling outlet.

Answer: (See page 643)

Question Type: S
Random: Y
Difficulty Level: 2
Chapter Number: 19
Page Reference: 643

89. Sketch the symbol used to show a floor outlet.

Answer: (See page 643)

Question Type: S
Random: Y
Difficulty Level: 2
Chapter Number: 19
Page Reference: 643

90. Sketch the symbol used to show a duplex outlet.

Answer: (See page 643)

Question Type: S
Random: Y
Difficulty Level: 2
Chapter Number: 19
Page Reference: 643

MULTIPLE CHOICE: Circle the letter indicating the best choice for each question.

91. Special outlets include:
  a. Entrance signals
  b. Telephone outlets
  c. Intercom systems
  d. Fire-alarms
  e. All of the above

Answer: e

Question Type: M
Random: Y
Difficulty Level: 2
Chapter Number: 19
Page Reference: 643

Chapters 20. DRAWING A SMALL COMMERCIAL BUILDING

1. List three general structural systems commonly used in commercial buildings.

Answer: Three general structural systems commonly used in commercial buildings are load bearing masonry walls, steel frame structures, and concrete frame structures.

Question Type: S
Random: Y
Difficulty Level: 2
Chapter Number: 20
Page Reference: xx

2. Why are reflected ceiling plans commonly drawn for commercial buildings?

Answer: Reflected ceiling plane are drawn for moat commercial buildings because an extensive amount of lighting and HVAC equipment must be mounted on the ceilings.

Question Type: S
Random: Y
Difficulty Level: 2
Chapter Number: 20
Page Reference: xx

3. When laying out plan views of buildings with multiple floor levels, why is it important for the drafter to reserve suitable space for vertical plumbing and mechanical chases?

Answer: Plumbing pipe, electrical wire, and HVAC duct work must be run vertically in a multi-story building; therefore it is necessary to plan for these mechanical services.

Question Type: S
Random: Y
Difficulty Level: 1
Chapter Number: 20
Page Reference: xx

4. What detail is commonly drawn before starting the plans and elevations?

Answer: The wall section detail is commonly drawn before starting the plans and elevations.

Question Type: S
Random: Y
Difficulty Level: 1
Chapter Number: 20
Page Reference: xx

5. What factors must be considered in deciding on the spacing of modular columns in the planning of a commercial building?

Answer: Limitations of the span of the beams framing the floor or roof structure would determine the spacing of modular columns in the planning of a commercial building.

Question Type: S
Random: Y
Difficulty Level: 2
Chapter Number: 20
Page Reference: xx

6. In dimensioning, how are features in detail views made to relate to major features on plan and elevation views?

Answer: There are more dimensions on detailed views than on plan and elevation views. such dimensions clarify information which cannot be shown on plans and elevations.

Question Type: S
Random: Y
Difficulty Level: 2
Chapter Number: 20
Page Reference: xx

7. Why must masonry coursing sizes be accurately drawn on details?

Answer: It is critical to draw masonry coursing sizes accurately on details so that the construction materials will fit properly.

Question Type: S
Random: Y
Difficulty Level: 2
Chapter Number: 20
Page Reference: xx

8. In a large commercial building the architectural firm completes all of the working drawings. True or False?

Answer: False

Question Type: T
Random: Y
Difficulty Level: 1
Chapter Number: 20
Page Reference: xx

9. Why do we have codes regulating the construction of commercial buildings?

Answer: We have codes to regulate the construction of commercial buildings to protect and insure the public health, safety and welfare.

Question Type: S
Random: Y
Difficulty Level: 1
Chapter Number: 20
Page Reference: xx

10. Discuss the purpose of the zoning regulations.

Answer: The zoning regulations govern the size of the property, the use of the property, the placement of the structure on the property, the size of the "footprint" of the building through the floor-area ratio (FAR) method, the height restrictions of the property, noise abatement requirements, the site open area requirements and parking requirements.

Question Type: E
Random: Y
Difficulty Level: 2
Chapter Number: 20
Page Reference: xx

11. How does the building code influence construction?

Answer: The building code influences construction by governing the use of the building, the fire ratings of the various materials used in the structure, space requirements, structural requirements, lighting requirements, HVAC requirements, sanitation requirements, access and egress and the safety of the general public when using the building.

Question Type: S
Random: Y
Difficulty Level: 3
Chapter Number: 20
Page Reference: xx

12. Name the eight general factors essential in planning a small commercial site.

1. ______________________
2. ______________________
3. ______________________
4. ______________________
5. ______________________
6. ______________________
7. ______________________
8. ______________________

Answer: The eight general factors essential in planning a small commercial site are:

1. Zoning regulations
2. Code requirements
3. Site/contextual analysis
4. Land utilization plan
5. Vehicular/pedestrian traffic
6. Parking facilities
7. Erosion control
8. Landscaping

Question Type: F
Random: Y
Difficulty Level: 2
Chapter Number: 20
Page Reference: xx

13. Acoustical control must be considered when designing a small commercial building. True or False?

Answer: True

Question Type: T
Random: Y
Difficulty Level: 1
Chapter Number: 20
Page Reference: xx